Jack S. Cohen

The Revolution in Science in America in the Twentieth Century

nova
science publishers

www.novapublishers.com

NOTICE TO THE READER

Library of Congress Cataloging-in-Publication Data

ISBN: 979-8-89113-967-1 (softcover)
ISBN: 979-8-89530-013-8 (e-book)

Published by Nova Science Publishers, Inc. † New Yor.k

Acknowledgments

I am greatly indebted to my son Simon Cohen, MS, who read this book and provided numerous corrections and suggestions. I thank Dr. Niklas Hebing and Mechthild Koehler, Librarian of the DFG, for their help.

Dear Kenny

What? Another book?

Contents

Preface

How did the US change from lagging behind Europe in science and technology at the beginning of the twentieth century, to being the super-power of science and technology by the end of that century? In attempting to answer this important question I first published an article on the subject [1]. The current book is an expansion and an improvement on this theme.

I can say with certainty that there is no other book that covers precisely this subject, which should fascinate every American. How America came from behind and overtook the rest of the world as a powerhouse in science and technology in the twentieth century is not a simple story, it involves a history of science and society that is neither linear nor straightforward. Yet every American, every housewife, every patient, every worker, indeed everyone on this planet has been intimately touched by the consequences of this revolution in science.

There were so many changes between 1900 and 2000 that it is hard to keep track. For example, in 1900, German was the language of science, but by the late twentieth century it had been supplanted by English. During the early twentieth century American doctoral students went to Europe to study, by the end of the century most Europeans went to the US to study.

War changed completely, the use of tanks and planes that were introduced in World War One became dominant in World War Two. The study of the atom that was obscure in the early part of the twentieth century, led to the atomic bomb and revolutionized the world. DNA that had been discovered in 1869, was completely obscure until its structure was determined in 1953 and then its analysis revolutionized forensics.

Just as the railway revolutionized travel in the nineteenth century, so the automobile and the Interstate Highway system revolutionized travel in the twentieth century. Then there were commercial airplanes and flights crisscrossing the country. And discoveries of the transistor and electronics and lasers and computers. And all mainly because the US Government and US industry finally after World War Two decided to fund basic research in all areas.

It was this realization, that the funding of basic research, without any clear application, that underlay most of these spectacular advances. That is the story I attempt to convey.

Chapter 1

Introduction

Americans are used to thinking of their country as the greatest in the world, both in terms of economic clout and military strength. But few know how the United States achieved this status. The fact is that the US became the greatest industrial power, out-performing the UK, its parent country, in industrial output (measured as GDP (Gross Domestic Product) per capita, to correct for different sized populations) by circa 1890 [2]. and has been estimated to have out-produced all of Europe around 1917, during World War One [3].

But in military terms the US had no "regular" Army as generally understood until 1913, when Secretary of War Henry L. Stimson organized one, in the form of four divisions assigned to protect each geographical region of the USA [4]. At this time, the UK had a large military force both fighting in and occupying colonies throughout the world. For example, at the Battle of Waterloo in 1813, the British Army, consisting of regular and conscripted forces, numbered around 120,000 men [5].

But, at least until World War One, and more generally until World War Two, the US was still a secondary power, especially in scientific terms. As late as 1999, Members of Congress were intent on proposing legislation to prevent contact between American scientists and foreign scientists and to reduce foreign students at American universities in order to protect American secrets [6]. Most of the great discoveries and basic research that revolutionized Western society were made in Europe, in the UK, Germany and France. But eventually the US out-stripped its European rivals in science too. How this happened is a unique and intriguing story.

The first organized attempts to improve US scientific standing was made in 1901 with the foundation of the Rockefeller Institute for Medical Research in New York City by John D. Rockefeller [7], and in 1903 with the formation of the Carnegie Institute of Washington (CIW), founded by Andrew Carnegie, the Scottish immigrant steel magnate [8]. He specifically envisaged that the CIW would engage in *basic research* (without specific applications) in all areas of science. But, over time, the CIW's impact was limited.

The next great attempt to expand American science was made during and after World War Two by Vannevar Bush, an extraordinary intellect, who envisaged an early version of the internet, and who was appointed Adviser for

Science and Development by President Roosevelt [9]. His influence caused a revolution in how science and basic research was thought of in America, both by the Government and by its people.

Most people would be shocked to discover that the US became the great scientific and technological power it is today by ironically exploiting two groups of Germans, first German (and other European) Jewish émigré scientists before World War Two and then German scientists, particularly German rocket and aeronautical engineers, after World War Two.

I endeavor to tell the story of how America became the world's scientific superpower through developments in science and technology.

Chapter 2

America Lags Behind Europe in Science at the Beginning of the Twentieth Century

1. Electricity Lights the New Century

Although the turn of the century was still seven years away, the World's Fair in 1893 in Chicago, also known as the World's Columbian Exposition, can be considered the event that presaged the advent of modernity. Not only was every conceivable new device on show there, the Bell telephone, the Singer sewing machine, the refrigerator, and so on, but it was the first major event to be lighted by electricity [10]. While the incandescent electric light bulb was an invention of Thomas Edison [11]. the electricity itself was the first major use of alternating current (AC) invented by his former assistant Nicola Tesla [12].

No-one could have imagined what incredible and amazing discoveries lay ahead that would revolutionize society and change people's lives.

Already, the telegraph that had been invented by Samuel Morse in 1844 had revolutionized long-distance communication. Thomas Edison invented an improved carbon transmitter for telephones in 1877 and the phonograph in 1878. But, it was the development of the first successful light bulb in 1879 that ensured he would become famous [11].

In search of a way to light up the city he formed the Edison electric light company in NY City to bring electricity to the masses. He said "We will make electricity so cheap that only the rich will burn candles" [13]. To do this he invested in what became known as Direct Current (DC) electricity, rather than the invention of one of his assistants, Nicola Tesla, an immigrant from Croatia, of Alternating Current (AC), a decision he would later regret. The competition between DC and AC has been described in detail [12]. AC was found to be by far the best for transmission over long distances and less dangerous. When it was chosen to light the World's Fair in 1893 and then the Pan-American Exposition in Chicago in 1895 [10], the stage was set for the electrification of America and subsequently the world. The future looked bright!

Notwithstanding these developments pioneered by Edison and other inventors, there was a problem in America that few people foresaw. As developments in science and technology proceeded at a rapid pace in the early years of the twentieth century, the US fell behind. Aside from the development of heavy industry, including steel production and extensive railway systems, and all the practical developments taking place in the US that made life easier for the average citizen, including typewriters, and all those devices dependent on electricity, such as, washing machines, dishwashers, tractors, and automobiles, these were all the result of *applied* research.

It should be pointed out that the first use of the term "basic research" was coined around 1920 in the US Department of Agriculture to mean development of improved means of crop production, what today we would classify as applied research. However, the term soon came to mean what had previously been called fundamental or pure research by scientists, since it was more readily understood by laymen [14].

America was great at the exploitation of new ideas with utility based on applied research, but *there was no organized attempt to foster basic research in America*. It was felt that industry should perform applied research to develop useful products and it would be a waste of taxpayer's money for the US government to support fundamental or basic research with no practical application in mind. By contrast in Europe, Germany and the UK had active and already traditional frameworks of fostering basic research at many famous universities and industrial laboratories.

2. Nobel Prizes Won

One can see using the ratio of Nobel Prizes as a convenient measure of scientific activity and success that America lagged behind the European nations at the beginning of the twentieth century (Figure 1) [15, 16]. Nobel Prizes were given for such important work in physics as the discovery of radioactivity, the nature of the electron and the atom, in chemistry for the development of dyes and drugs, in physiology for the understanding of hemoglobin and the function of proteins and enzymes. In all these areas the research level and competition were much more intense in Europe than in America at that time.

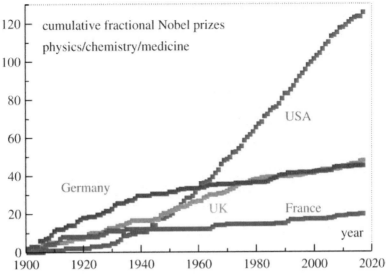

Source: Permission granted by University of Firenze Press.

Figure 1. The cumulative number of physics, chemistry and medicine Nobel prizes per country. Prizes are attributed to the respective country according to the nationality of the recipients at the time of the announcement, with prizes obtained by more than one recipient accordingly divided. Note that the US population increased from 76 to 327 million during 1901–2017 [15].

Here is a partial list of some early German Nobel Prize winners: In Physics; Roentgen (1901), Lenard (1905), von Laue (1914), Planck (1918), Stark (1919), Einstein (1921), Hertz (1925), Franck (1925), Heisenberg (1932). In Chemistry; Fischer (1902), von Baeyer (1905), Buchner (1907), Ostwald (1909), Wallach (1910), Willstatter (1915), Nernst (1920), Wieland (1927), Fischer (1930), Bosch (1931). These names are of the highest possible caliber and were responsible for significant achievements in numerous scientific subjects at the time.

By comparison during this period the US had the following Nobel Prize winners: Physics; Michelson (German Jewish immigrant, 1907), Millikan (1923), Compton (1927), Davisson (1937), Lawrence (1939); Chemistry, Richards (1914), Langmuir (1932), Urey (1934). Frankly, there is no comparison in quantity or accomplishments. The same could be said of a comparison of UK and US Nobel Prize winners during the same period 1901-1939.

3. Britain and Germany Are Ahead in Research

What is the origin of these differences in Nobel Prize achievements? In the UK, the Royal Society, the oldest such scientific organization, was founded in 1660. It supported a culture of research, including such illustrious members as Sir Christopher Wren (architect who built St. Paul's Cathedral), Sir Isaac Newton (author of *Principia Mathematica*), Robert Boyle (who discovered the first law of physics), and so on. Government funding of science started in 1675 when the Royal Observatory was established in Greenwich. This was continued in the 19th century with the creation of the British Geological Survey in 1832, and the allocation of funds in 1850 to the Royal Society to award individual grants [17].

In Berlin in 1909, Professor Adolf von Harnack, a close adviser to Kaiser Wilhelm II and a member of the Academy of Sciences, wrote a memorandum to the Kaiser in which he outlined a reform of the German science system. He proposed the establishment of independent research institutes conducting specialized basic research. He wrote that the rapid pace of industrialization had demonstrated the need for greater knowledge of basic sciences. Harnack proposed the foundation of a new type of research association for the advancement of science to be known as The Kaiser Wilhelm Society. Harnack's memorandum paved the way for a reorganization and the establishment of research Institutes that still characterize the German science system today [18].

The Kaiser Wilhelm Gesellschaft (KWG) was founded in 1911 for the advancement of science and was formally independent of the German state. Some 30 Research Institutes and testing stations were founded all over Germany in specific areas of science! The KWG had Presidents such as Adolf von Harnack, Fritz Haber, Otto Hahn and Max Planck, and each Institute had its own Scientific Director. Funding was obtained from inside and outside Germany.

By comparison to the European powers, the fact is that in the US at the beginning of the twentieth century there was *no* Federal Government support for basic research, there were no institutions that were funded to carry out basic research and no committees existed to foster such research. It was customary for American students to go to Europe to obtain advanced degrees before they could expect to receive a position in a university or company in the US. It is no wonder that the US fell behind Europe in the early part of the twentieth century in the advancement of *basic* science and the advantages that it provides for industry and society.

Chapter 3

Two Robber Barons Establish Research Institutes in America

1. First Medical Research Institutes in America

The origin of the National Institutes of Health (NIH) can be traced to the Marine Hospital Service started in the late 1790s that provided medical relief to sick and disabled men in the U.S. Navy. By the 1870's a network of Marine Hospitals was developed and Congress allocated funds to investigate the causes of epidemics like cholera and yellow fever. The National Board of Health was also created, making medical research an official government initiative [19].

Source: Public domain image, NIAID/NIH.

Figure 2. The Hygienic Laboratory ca. 1890.

In 1887, the US government established the Hygienic Laboratory at the Marine Hospital on Staten Island, New York. The laboratory was tasked with applying the new science of bacteriology to the study and diagnosis of epidemic diseases (Figure 2).

Joseph James Kinyoun, a young scientist and physician, was the first employee. His work involved collecting blood and stool samples from sick patients to culture pathogens in the lab.

Kinyoun (Figure 3) was able to isolate the microorganism that caused cholera, a disease that had killed hundreds of thousands [20]. Kinyoun visited Europe several times to train with renowned bacteriologists Robert Koch, Louis Pasteur and others and brought back techniques and treatments to reform US health practices. Kinyoun studied such infectious diseases, as cholera, yellow fever, smallpox, and bubonic plague. The Marine Hospital Service was responsible for diagnosing infectious diseases among immigrant passengers on incoming ships to prevent disease from entering the US. Congress recognized the value of the laboratory, enabling it to expand.

Source: Public domain image, National Library of Medicine.

Figure 3. Joseph. J Kinyoun.

In 1891 the Hygienic Laboratory moved to larger quarters in Washington DC. As a result of the passage of the Biologics Control Act in 1892, the Laboratory was now responsible for the regulation of vaccines and other biologic products. In 1918 the laboratory combatted the Spanish influenza epidemic and cared for veterans of World War One In 1942 the Hygienic Laboratory transferred to Bethesda, Maryland, upon a gift of land from the Stone family to the US government, where it grew into what is today the National Institutes of Health, the largest biomedical research Institute in the world employing today over 18,000 people (Figure 4) [19].

Imagine living in the past when plagues and pandemics killed tens of thousands or more people and you had no idea why, where it came from and how it worked. During the middle ages the bubonic plague known as The Black Death starting in 1347 killed *one third of the population of Europe*. In 1900,an epidemic of scarlet fever swept the US, causing the random deaths of many thousands of children and adults.

Source: Public domain image, NIH.

Figure 4. The campus of the National Institutes of Health (NIH) in Bethesda Maryland, USA, consisting of many disease-related research Institutes. The Clinical Center is in the foreground.

One of those who died in this epidemic was the grandson of the immensely wealthy industrialist John D. Rockefeller. Despite his wealth Rockefeller was unable to save his grandson, John Rockefeller McCormick, named for him. His mother Edith McCormick was of course grief stricken. His death had significant consequences, his mother was galvanized to take action.

In 1902, Edith and her husband founded the John R. McCormick Memorial Institute for Infectious Diseases in his name in Chicago. The articles of incorporation explicitly stated that the Institute was established for "the study and treatment of scarlet fever and other acute infectious diseases."

The Institute was established as a two-story building with 60 beds with an adjacent laboratory. The hospital provided free care to patients, and the research lab's primary goal was to develop and distribute vaccines. In 1914, Gladys and George Dick joined the Institute, with their focus on solving the cause of scarlet fever. Gladys Dick had graduated from Johns Hopkins Medical School and studied biomedical research while in Berlin. It took almost a decade for the Dicks to finally discover the cause of scarlet fever. The bacterium they discovered became the object of a diagnostic skin test and a vaccine for treatment. An immunization protocol soon followed, drastically changing the survival rate for scarlet fever patients.

Meanwhile John D. Rockefeller himself, perhaps motivated by his daughter's example, had decided to found an even larger Institute for the study of infectious diseases. It should be noted that he had also lost a baby daughter, Alice, to dysentery in 1870. In deciding to found an Institute he was perhaps also motivated by the example of other immensely wealthy "robber barons" who had or were in the process of establishing eponymous Institutes. One of these was Andrew Carnegie who was discussing founding an Institute dedicated to basic research.

2. The Development of the Germ Theory of Disease

What was it fundamentally that led Rockefeller to establish his Institute for Medical Research? Although his family losses no doubt played a role in his decision, in reality it was a series of discoveries that occurred throughout Europe that led to the development of the "germ theory of disease." In other words, the realization that infectious diseases in humans are caused and promulgated by germs, microorganisms, and once that was realized prevention and therapies could be developed.

The first step in this process was the invention of the microscope around 1590 by lens makers Hans and Zacharias Janssen in Holland, who created a microscope based on lenses in a tube. No observations from their microscopes were published and it was not until Robert Hooke in England and Antonie van Leeuwenhoek in Holland improved the magnification and published their results that the microscope, as a scientific instrument, was born. Particularly Hooke's publication *Micrographia* in 1667 of his observations of microorganisms was a key event. Leeuwenhoek then described his observations in 1675 and what is relevant for our subject is that he was the first to observe bacteria [21].

Originally scientists and physicians believed in the spontaneous generation if disease, as if it appeared out of thin air. Realizing that microorganisms, and specifically bacteria were the culprits came after much investigation. It was Louis Pasteur in France in 1862 who convincingly proved this. He conducted a series of experiments whereby he boiled broth to destroy all possible microorganisms in a series of glass vessels which were sealed with a water trap. No growths were detected in the broth, but when he broke the seal, within 3 days growths appeared [22]. It was then clear that microorganisms in the air were the cause of these growths and ultimately of human infectious diseases.

It then became the goal of medical scientists to investigate the fluids from patients to identity, using microscopy, which bacteria were responsible for which diseases. Then how to prevent the spread of disease by killing them. Pasteur discovered the pneumococcus bacterium, the cause of pneumonia. In Norway, Gerhard Hansen discovered the bacterium responsible for leprosy. Robert Koch, a German physician, found the bacteria that causes tuberculosis, for which he was awarded the Noble Prize in Medicine in 1905 [23]. In their honor the Pasteur Institute for Medical Research was founded in Paris in 1897 and the Robert Koch Institute was named after him in Berlin in 1899. In a very real sense the McClintock and the Rockefeller Institutes were founded based in part on the prior foundation of the Pasteur and Koch Institutes

3. The Founding of the Rockefeller Institute for Medical Research

It is not unusual for people of immense wealth to give money to charity. And it was not unusual for such people as the industrialist Andrew Carnegie to donate more than $40 million to build over 1,600 public libraries throughout America from 1896. But, the foundation of Institutes to carry our research in basic and medical sciences is quite unique.

In the US at the turn of the century, many so-called "robber barons" had made huge fortunes in such industries as iron, steel, coal, railways and automobiles [24]. Several of them in later life turned to philanthropy and established Institutes in their name. Thus there is the Frick Institute on the Mall in Washington DC, established by Henry Clay Frick, who made his fortune in steel and railroads, that holds a wonderful art collection. Leland Stanford in California, who made his fortune in railroads, chose to establish a famous University.

Two of these luminaries chose to establish Institutions that support the concept of basic research, John D. Rockefeller (Figure 5), who made his fortune in oil refining and was reputed to be the wealthiest American, established the Rockefeller Institute for Medical Research in New York City in 1901 [7]. Andrew Carnegie (Figure 6), who had made his fortune in iron and steel production, realized the need for *basic research* in America and founded the Carnegie Institute of Washington (CIW) [25]. No doubt there was a degree of competition between these two industrial giants.

Source: Public domain, image from 1907.

Figure 5. John D. Rockefeller.

Source: Public domain, Library of Congress.

Figure 6. Andrew Carnegie.

On the advice of his friend Frederick T. Gates and the influence of his son Rockefeller initially donated $200,000 for the establishment of his Institute. He assembled a group of friends and experts to form the Board of Directors of the putative Institute, and they met in 1901 to initiate this venture. After many considerations the Institute was founded in New York City in 1901, where the board selected one of their members Simon Flexner to be the Director.

Source: Public domain, National Library of Medicine, 1937.

Figure 7. Oswald T. Avery, who proved that DNA is the genetic substance.

Among the many advances that were to be discovered at this prestigious Institute, perhaps the most important was the significance of DNA as the genetic substance. Although DNA was discovered by Friedrich Miescher in 1869 in Tubingen, Germany, it was considered an obscure and unimportant substance for almost a century [26]. Scientists believed that proteins must be the basis of the genetic information that underlay life, partly because there were 20 amino acids in proteins, but only 4 bases in DNA.

At the Rockefeller Institute there were three individuals who worked on DNA, Phoebus Aaron Levene, a Jewish immigrant from Russia who worked on the chemical structure of DNA, Alfred Mirsky worked on the proteins that bound to DNA, and Oswald T. Avery (Figure 7) studied the function of the pneumococcus bacteria.

It was known that there were two forms of this bacteria, one that was pathogenic, in other words it caused the disease, and one that was nonvirulent. In a series of elegant experiments reported in 1944, Avery and his colleagues showed that when extracts from the virulent form were transferred to the nonvirulent form only the DNA extract caused it to become virulent, while the protein extract did not.

In doing this he *had proved that DNA was the genetic substance*. Mirsky still preferred to believe that proteins were more important than DNA, and since Avery was already nearly 67 years old when he did this work and soon retired, it took further discoveries before the scientific world became convinced of the genetic significance of DNA [26].

4. The Founding of the Carnegie Institute of Washington

Carnegie's initial donation of m$10 for the purpose of founding his Institute was given with the stipulation that only research without any applied objectives should be conducted there. He hoped that this would engender a commitment to basic research throughout America.

Once established in 1903, the CIW engaged in many areas of research, including physics, chemistry, genetics and astronomy. This included Edwin Hubble who revolutionized astronomy in 1929 with his discovery that the universe is expanding, and Barbara McClintock, who won the Nobel Prize in 1983 for her work on genetics in maize.

Although the CIW did make important contributions in many areas it is interesting to note that while the first Director of the CIW, Robert S. Woodward, was himself a physicist, the area of nuclear physics was completely ignored. Under the influence of the Board member Henry S. Pritchett, who served as Director of the U.S. Coast and Geodesic Survey, the major project in physics that the CIW undertook to pursue was the construction of a wooden-copper boat *The Carnegie*, to sail the seas of the world and determine the earth's magnetic field [27]. This could not be done obviously in a regular iron ship. But unfortunately, the ship was destroyed by fire when it was docked in Tahiti.

But the CIW missed the boat as it were in physics, they chose not to work on the frontline in physics research that was taking place in Europe, where such notables as Rutherford in Britain, Neils Bohr in Denmark, Werner Heisenberg and Albert Einstein in Germany, were grappling with the structure of the atom and its properties. If they had initiated a program of research into the atom, the US might not have needed to depend on the immigration of European Jewish scientists in the 1930's to initiate the Manhattan Project to build an A-bomb (see Chapter 6).

Although the CIW did some notable basic research, its influence was not so great as to bring America in line with its European competitors. At the time German was considered the scientific language. Andrew Carnegie's hope that CIW would bring about a revolution in support for basic science in America was not realized.

Chapter 4

Lessons of World War One for Science and Technology

1. World War One and the Development of New Weapons

It is a well-known truism that warfare results in advances in science and technology that have tremendous long-term consequences. This is certainly true of the many developments in military technology that resulted from World War One.

The development of the tank is an excellent example. Originally ideas of a mobile armored vehicle were conceived by Leonardo da Vinci and H. G. Wells. Wells called them "Land Ironclads" and described their successful use in his story of that name published originally in 1903 [28]. But, in reality it took a lot longer to develop them. At first an American company named Holt of Stockton CA developed caterpillar tractors that were used to tow heavy equipment around behind the lines of the UK forces in World War One.

One of the first such practical designs was the "landship" proposed by British engineer and inventor, Sir Ernest Dunlop Swinton. Swinton's design was based on a tractor that was armored and could move across the battlefield. The vehicle was equipped with machine guns and was designed to break through enemy lines. However, Swinton's design was rejected by the British Army. It was not until 1915, when the situation on the Western Front became dire, that the British War Office began to seriously consider the idea of an armored vehicle. In that year, the British formed the Landships Committee, a group of military officers, and engineers who were tasked with developing an armored vehicle that could break through the trenches. The committee invited several engineers to submit designs, and in 1916, a prototype was built by William Foster & Co., an agricultural machinery manufacturer. The vehicle was called "Little Willie" and was the first tracked armored vehicle [29].

The first time tanks were used in battle was during the Battle of the Somme in 1916. The British deployed 32 Mark I tanks, and while their initial success was limited due to mechanical issues and difficult terrain, the psychological impact on the Germans was enormous (Figure 8). The Germans had never seen anything like the tanks before and were terrified by their appearance and firepower. The tanks played a significant role in breaking the deadlock at the Somme and paved the way for the development of more advanced tanks in the following years. Lieutenant Basil Henriques, scion of a prominent Anglo-Sephardic-Jewish family [30], led the first tank column into battle at Cambrai in 1917. The actual result was a disaster, the tanks were not bulletproof and were picked off by the powerful German artillery, and they also got stuck in terrain. Nevertheless, it was hailed as a success by British propaganda and so the myth of the tank was born.

Tanks were first used effectively by Col. George Patton in the Battle of Amiens under the command of US General Pershing, that was effectively the last major battle of World War One [31].

During the 1920's many improvements were proposed, but mostly rejected by traditionally thinking military commands. However, one innovation that proved significant was that of independent suspension of all wheels of the track proposed by an American transportation engineer named J. Walter Christie in 1928 [32]. This allowed the tank to move much faster over rough terrain, precisely what a tank was needed to do. This idea too was rejected by US and British Army ordnance officials. But, the Germans, who had been defeated at Amiens by tanks, realized their military significance and took up this idea and incorporated it into their Panzer tanks.

This was one of the main reasons for the defeat of the French and the British Expeditionary Force at Dunkirk at the beginning of World War Two. They were stunned by the speed with which the German tank corps raced ahead and overpowered them. The Russians too took this idea and incorporated it into their tanks and eventually the Americans and British followed suit. Incidentally, one reason why Germany lost World War Two was that although they produced the "best" tank, the famed Tiger tank, they were over-engineered and were so heavy that they had to stop to fire, and Germany produced only 1,350 of these, while the US produced 49,324 Sherman tanks, that were more mobile, more easily repaired and less expensive [33].

Along with the development of the tank, during World War One, parallel advances were made in the areas of guns, airplanes, submarines and radio communication.

American-invented machine guns were increasingly used in World War One. Army Col. Isaac Newton Lewis invented the Lewis Machine Gun in 1911. Initially there was minimal interest within the U.S., but a few years later, the newly-retired officer showed his weapon to European buyers, who expressed great interest. Lewis had a factory built in Belgium, and approximately 100,000 automatic weapons saw service in World War One.

Only after the British and French had bought these weapons did America decide to purchase them. Lewis, already a wealthy man, declined the royalties on guns made for the United States after the country entered the war.

Benjamin Hotchkiss, an American ordnance engineer, also invented a machine gun that bears his name. The French infantry selected the Hotchkiss as its principal machine gun during World War One. The American Expeditionary Force in France followed suit, becoming the second-largest user of the Hotchkiss.

From the Wright Brothers first flight in 1903 in North Carolina, World War One catapulted the airplane into prominence, first for reconnaissance of enemy positions and then as fighters facing each other. In 1908, the U.S. Army Signal Corps sought competitive bids for a two-seat observation aircraft. The 1909 Wright Military Flyer was the world's first military airplane.

But in World War One America depended at first on Europe for aircraft. America used a plane designed by famed British aviation engineer Geoffrey De Havilland that had first flown in combat with the British Royal Flying Corps in 1917. When the United States joined the war in April of that year, an example was sent to the United States to determine if it was suitable for American production. The American Aircraft Production Board approved the design for construction in July 1917, pairing it with an American-designed V-12 "Liberty" engine. For this reason, American DH-4 planes were known as "Liberty planes." Thanks to an incredibly fast production schedule, the first Liberty plane rolled out of an American factory in October 1917.

Source: Public domain, UK Government image.

Figure 8. British Tank in World War One.

In the earliest scout planes, an observer leaned over the side holding a camera and took a photo, then had to change the glass-plate negative. The quality of the imagery was good, but it was a cumbersome process. Beginning in 1915, the British and French started to use cameras to photograph the German front. George Eastman, founder of the Eastman Kodak Company, wanted to help the country in this time of peril. He had the scientists and the facilities to establish an aerial photography school during the War. At first, the government showed no interest, but Eastman persisted. The government reconsidered, and in 1918, the last year of the war, the school opened in Rochester, New York to teach aerial photo-processing and camera repair.

2. Fritz Haber and the Development of Gas Warfare

Fritz Haber, a German Jew who won a Nobel Prize in 1918 for fixing nitrogen from the air, thus synthesizing ammonia in large quantities, also prepared the chlorine gas that Germany unleashed in trench warfare and that forced Allied soldiers to carry gas masks for protection.

Fritz Haber was influential and helped Albert Einstein secure a position at the University of Berlin and the Prussian Academy of Sciences in 1913. They were friends, but they fell out over Haber's support for German involvement in World War One, as opposed to Einstein's pacifism and opposition to the War [34]. Haber was a German nationalist, his Jewish forebears had been among the first to be accepted as citizens of Prussia, and they had prospered in Germany for generations [35].

Haber enthusiastically supported the war effort and was appointed Captain in the German army and Head of the Chemistry Section of the Ministry of War. He not only developed the production of chlorine gas for warfare, but was present at Ypres supervising its first use in 1917, with a team of 150 chemists and 1,300 technicians that resulted in 67,000 enemy casualties. He also helped develop gas masks to protect German soldiers against the gas [35].

Haber was widely criticized for his support of the use of chemical weapons, which killed and disfigured thousands of soldiers He rejected this criticism:

"The disapproval that the knight had for the man with the firearm is repeated in the soldier who shoots with steel bullets towards the man who confronts him with chemical weapons. [...] The gas weapons are not at

all more cruelthan the flying iron pieces; on the contrary, the fraction of
fatal gas diseases is comparatively smaller, the mutilations are missing"
[36].

However, his support for the use of poison gas was to have terrible
personal consequences, both his wife Clara, a very sensitive and vivacious
woman, and his daughter Claire committed suicide, although the reasons were
not expressed.

In 1933, after the Nazis came to power, Haber being Jewish was expelled
from all his positions and forced to emigrate from Germany. He spent several
months in Cambridge, where he met Chaim Weizmann, also a chemist and
leader of the Zionist movement. Weizmann offered Haber the position of Head
of the Sieff Institute (later the Weizmann Institute) in Rehovot, then in British
Palestine. However, Haber's health was worsening, and he died *en route* in
1934 while in a hotel in Basel.

Haber was one of the most important chemists in history. His process of
catalytic conversion of nitrogen from the air and hydrogen from the
electrolysis of water into ammonia with Carl Bosch, resulted in the efficient
production of explosives and fertilizer. Hundreds of thousands of tons of
fertilizer are produced annually by this process and without them the
population of the world would starve.

The weaponization of debilitating nerve gases came subsequently with
lethal phosgene introduced by the Germans, and in 1917 incapacitating
mustard gas proved exceedingly effective in breaking the deadlock of trench
warfare - but it was not as decisive as anticipated. The war gases produced far
fewer fatalities than other weapons. America also developed and used
chemical weapons. Future president Harry S. Truman served as the captain of
a U.S. field artillery unit that fired poison gas against the Germans in 1918.

3. Communications in World War One

At the turn of the 20th century, the world was being introduced to an array of
new technology utilizing electricity. In addition to the telegraph, the telephone
also played a significant role in World War One communications. Telephone
lines were laid out across the battlefields, allowing commanders to
communicate with their troops in real-time. The telephone was especially
useful for coordinating artillery fire and relaying orders to troops on the
ground.

Wireless telegraphy, also known as radio, was another technology that played an important role in World War One communications. It allowed for communication without the need for physical wires, which was particularly useful in situations where wires had been cut or destroyed. Wireless telegraphy was used extensively by the navies of both sides, allowing ships to communicate with each other and with their home ports.

Carrier pigeons and dispatch riders were also still used as a means of communication during World War One. Carrier pigeons were particularly useful in situations where other forms of communication were unavailable, such as during trench warfare. Dispatch riders, on the other hand, were used to physically transport messages between units or to headquarters.

Overall, the introduction of electrical communication technologies such as the telegraph, telephone, and wireless radio had a significant impact on the conduct of warfare in World War One. These technologies allowed for faster and more efficient communication, which in turn led to better coordination and decision-making on the battlefield.

4. Naval Warfare in World War One

The British fleet was approximately twice the size of the German fleet and managed to blockade Germany and helped bring about the eventual German defeat. Naval warfare was dramatic, but had the expected outcome with Britain having such numeric superiority over the Germans. The major clash between the two fleets occurred at the Battle of Jutland in May 1916 [37]. The German fleet of 99 ships under Admiral Reinhard Scheer sailed north in formation with the intention of making a surprise attack on the British fleet. But they were detected and they did not expect the British Admiral Sir John Jellicoe to commit his whole fleet of 151 ships to the battle. The British fleet blocked the German fleet and there were losses on both sides, the British lost three major ships during the engagement. However, the German fleet was ordered to about turn and they disappeared into the mist and returned to their port at Wilhelmshaven. Both sides were mainly concerned about conserving their fleets rather than risk losing them in further battles.

As part of the armistice negotiations in 1918, the German fleet was forced to surrender to the British and was escorted to the British naval base of Scapa Flow north of Scotland. There, rather than give their fleet intact to the British, the Germans under orders scuppered their fleet by blowing it up and escaping in small boats. Recently there has been interest in documenting the whole German fleet sitting on the sea bed in Scapa Flow.

As a result of the reluctance of either side to commit its major naval vessels to further combat, the emphasis in naval warfare focused on submarines [38]. Everyone knows that the German U-boats played a significant role in World War Two, but most do not realize that they also played such a role in World War One. World War One saw important advancements in submarine technology, including the development of the first diesel-powered submarines. These submarines were faster and more reliable than previous designs, but the Germans only possessed 28 of them when the War started in 1914 [39].

They soon expanded their fleet and at first started to attack British naval vessels. But soon they changed their strategy and switched to attacking and sinking merchant ships that the German Admirals realized could cripple British society. However this proved a mistake, since many of the merchant vessels were from neutral countries supplying Britain, such as the United States. The sinking of US passenger and merchant ships brought the US into the War against Germany and this proved to be crucial, and facilitated German defeat.

The US Navy at first had very little investment in submarines and even knowledge of how to use them. In the US Naval College at Annapolis a young officer named Hyman Rickover, a Jewish American who spoke German, was asked to translate the crucial book, *Das Unterseeboot* (The Submarine) by Admiral Hermann Bauer that was the bible of German submarine practice. The translated book was a secret as Americans learned how to build and use submarines and counter-attack German submarines. Rickover went on to become an Admiral and famously develop the first atomic-powered submarines [40].

5. Development of Air Transport

In the early 1920s, the United States fell behind Europe in the development of air transport. Despite the restrictions imposed by the Treaty of Versailles, the Germans displayed exceptional ingenuity by designing high-quality aircraft and operating them in foreign countries. A notable venture took place in South America when a group of Colombian and German businessmen founded the Sociedad Colombo-Alemana de Transportes Aéreos (SCADTA) on December 5, 1919 [41].

SCADTA demonstrated remarkable initiative by importing Junkers-F 13 aircraft with floats, which were used to operate along the Magdalena River. The airline commenced operations on September 19, 1921, and has been in service ever since. Today, AVIANCA, Colombia's national airline, is a direct descendant of SCADTA and holds the distinction of being the oldest airline in the Americas.

Under the leadership of Peter Paul von Bauer, the head of SCADTA, the company expanded its domestic network in Colombia and set its sights on North America. Fritz Hammer, a skilled German salesman representing the Condor Syndikat, arranged for two Dornier flying boats to be shipped to Colombia. These aircraft were owned by Condor, but were leased to SCADTA.

Von Bauer embarked on an audacious venture by leading a delegation to the United States in the two Dorniers. They departed from Barranquilla, Colombia, in August 1925 and arrived in Havana. However, the U.S. response was hesitant, and after some deliberation only one plane was permitted to fly to Florida. While Hammer went to New York to seek business support, von Bauer traveled to Washington, D.C., and met with President Coolidge. Unfortunately, his efforts yielded little result, with only the Commerce Department showing genuine interest.

Frustrated by the unsuccessful negotiations, the delegation returned to Colombia. If the talks had been successful, SCADTA could have launched a trans-Caribbean service, significantly altering the course of airline history. But Von Bauer's expedition certainly sparked interest in the United States. Within a few weeks, on January 8, 1926, the State Department organized an interdepartmental conference. Shortly thereafter, the Air Commerce Act and the Foreign Air Mail Act were enacted. The United States was on the verge of entering the international commercial airline arena; all it lacked was an airline to fulfill that role.

Established in 1927 by two former U.S. Army Air Corps officers, Henry 'Hap' Arnold and Carl Spaatz, Pan American Airways or Pan Am initially operated as a mail and passenger service between Key West, Florida, and Havana, Cuba. In the 1930s, under the leadership of American entrepreneur Juan Trippe, the airline acquired a fleet of flying boats and expanded its routes to Central and South America. Over time, Pan Am ventured into transatlantic and transpacific travel, solidifying its dominance in international aviation. It played a pivotal role in the Jet Age, procuring modern jetliners like the Boeing 707 and Boeing 747. With its advanced fleet, Pan Am offered increased capacity, longer flights, and fewer layovers compared to its competitors. The airline's primary hub and distinctive terminal was located in New York City's John F. Kennedy International Airport.

From the late 1950s to the early 1970s, Pan Am enjoyed a prestigious reputation worldwide due to its cutting-edge fleet, highly skilled personnel, and exceptional amenities. In 1970 alone, it transported 11 million passengers to 86 countries across all continents. Operating in an era when most flag carriers were government-owned, Pan Am became the *de facto* national airline of the United States.

However, starting in the mid-1970s, Pan Am confronted a series of internal and external challenges, including increased competition following the deregulation of the airline industry in 1978. Despite several attempts at financial restructuring and rebranding throughout the 1980s, Pan Am gradually sold its assets and eventually filed for bankruptcy in 1991 [41].

6. British Initiatives in Science after World War One

By 1915, during the First World War, claims about the poor state of British manufacturing compared to Germany, led to the founding of the Department of Scientific and Industrial Research (DSIR) in the UK. It was a part of the UK government, staffed by civil servants who distributed grants, operated laboratories, and made policy. Examples included the Radio Research Station, established in Ditton Park in 1924.

In 1918, Richard Haldane produced an official report on the machinery of government that recommended that government departments undertake more research before making policy [42]. It was recommended that they should oversee that specific, policy-minded research was carried out, governed by autonomous councils free from political pressure.

Following the Haldane Report's recommendations, the Medical Research Council (MRC) was created in 1920 from a previous body called the Medical Research Committee that had been established in 1913 to distribute funds collected under the National Insurance Act of 1911. In contrast to DSIR, the MRC was not a government department, its staff were not civil servants, and its resources were concentrated in a small number of central laboratories and a large number of research units associated with universities and hospitals [17]. This is still the pattern today.

The initiatives undertaken in this period laid the groundwork for the development of a strong scientific infrastructure in the UK. The DSIR's role in promoting industrial and technological innovation, along with the MRC's focus on medical research, led to enduring impacts on various fields, including communication technology, healthcare, and scientific policy.

7. Initiatives in Science in Germany after World War One

After the First World War the financial situation of the universities and scientific institutions in Germany was dire. Their budgets had not been increased since before the War and inflation was rapidly increasing. However, it was precisely in this period following the War that an increase in funding was most needed. The War had been responsible for the interruption of scientific and research activities, young researchers had been called up for military service and research projects had been interrupted. In addition, basic research had been almost completely discontinued in favor of research critical to the war. This situation was further exacerbated by the international isolation of German research, because of the Treaty of Versailles, which ascribed sole guilt to Germany for the First World War.

In 1920, leading representatives of science and scholarship in Germany established a working committee, which subsequently adopted the name Notgemeinschaft ("emergency foundation"). Its task was to coordinate joint action and proposals to the parliaments, governments and also potential sponsors in industry, in order to secure the provision of the necessary financial resources to pursue *basic* research.

Friedrich Schmidt-Ott, Adolf von Harnack and Fritz Haber (Figure 9) played leading roles in this working committee, and also in lobbying the government for research funding [43, 44]. Friedrich Schmidt-Ott was elected president of the Notgemeinschaft at the inaugural meeting in 1920. Adolf von Harnack was the President of the Kaiser Wilhelm Society (KWG, later the Max Planck Institutes) founded in 1911. Fritz Haber was director of the Kaiser Wilhelm Institute of Physical Chemistry and Electrochemistry in Dahlem. He and Adolf von Harnack became members of the Executive Committee of the Notgemeinschaft in 1920.

The concerns of the Notgemeinschaft fell on sympathetic ears in government and in society; a decline in the standard of German research compared with other nations was seen as a loss of national honor. In addition, there were concerns about a negative impact on Germany's future economic development. In an application by the Notgemeinschaft for financial support from the Reich Government in 1920, Adolf von Harnack stressed the importance of science and research for Germany's overall development:

"The vital necessities of the nation include the preservation of the few assets that it still possesses. Among these assets, German science and research occupy a prominent position. They are the most important prerequisite not only for the preservation of education in the nation and for Germany's technology and industry, but also for Germany's reputation and its position in the world, on which in turn prestige and credit rely."

Source: Reproduced by permission, University of Firenze Press.

Figure 9. *Left*, Friedrich Schmidt-Ott, President of the Notgemeinschaft from 1920 to 1934; *Middle*, Adolf von Harnack ; *Right*, Fritz Haber.

Following debates on the allocation of Reich funds to the nascent Notgemeinschaft in the Reichstag [45], in October 1920 the Reich Ministry of the Interior made 20 million marks available in the 1921 budget year "for the advancement of the goals pursued by the Notgemeinschaft der Deutschen Wissenschaft." Funding continued in this manner until 1934, when the Committee of the Notgemeinschaft were forced to resign and were replaced by Nazi Party control. Haber who was born Jewish, had converted to Christianity and was a German nationalist, was nevertheless dismissed from all his positions and left Germany and died in poverty in Basle, Switzerland in 1934.

The Notgemeinschaft was the precursor of the Deutsch Forschungsgemeinschaft (DFG) the German Research Foundation after World War Two that was founded officially in 1951 and became the Federal organization for the support of basic research in the Federal Republic of Germany. Also after World War Two the Kaiser Wilhelm Institutes became the Max Planck Institutes [46].

Chapter 5

The Dark Side of America

1. Racism and Violence in America

It would be a disservice to the reader and inaccurate to foster the belief that America was "the white city on the hill," that there was no downside and that science flourished there because that is what all Americans wanted. That is far from the truth. While the US may have been a country "conceived in liberty," nevertheless it was also a country based on genocide where racism and violence were rife.

How else to describe the treatment of the native peoples except as genocide. This was a clash of civilizations in which the indigenous peoples were doomed to lose [47]. As the white man poured across the plains they murdered "Indians," they put their children in schools more like prisons, often administered by Churches, where they were forbidden to speak their own native languages and were beaten and murdered [48].

Then there was the "trail of tears" when the US Army forcibly removed the Cherokee, Cree, Choctaw and other nations comprising some 60,000 people from Georgia to Oklahoma, a distance of 860 miles, in 1850-52, a clear case of ethnic cleansing [49]. Oklahoma means "red man's land" and the US government promised it to the tribes as Indian Territory "in perpetuity" in 1834, a period that actually lasted until 1907, when Oklahoma became a US State.

How could a country in which democracy was regarded as the ultimate goal of civilization have decided that black people, had no rights? In the famous Dred Scott case of 1857 the US Supreme Court ruled that Black people are not included and were not intended to be included under the word 'citizen' in the Constitution, and so were not afforded the rights and privileges that it granted to American citizens [50]. Racism and lynching flourished and violence was the norm. For example, in 1919 there were anti-black race riots in Wilmington, Delaware, and the federal government refused to intervene [51].

Add to this mixture a good dose of anti-Semitism, including the lynching of Leo Frank in Marietta, Georgia, for the murder of Mary Phagan, of which he was certainly innocent (he was posthumously pardoned in 1986) [52]. As well as prevalent anti-Semitism in the State Department that prevented Jews immigrating to America during the rise of Nazism in Germany (see Chapter 6). During the 1920's-30's it was estimated that 40% of white American males belonged to the Ku Klux Klan, the John Birch Society or the German-America Bund. It is indeed a miracle that out of this cauldron of excesses and racism, liberty, democracy and science prevailed.

2. The Eugenics Movement in America

Against this backdrop it is not then surprising that science was harnessed as an excuse to justify racism and repression of minorities. The pseudo-science of *eugenics* thrived for some time, and was used to rid society of undesirables, the poor, the disabled, the mentally ill, and specific communities of color by forced sterilization [53].

The American eugenics movement emerged in the late nineteenth century and was influenced by the ideas of Sir Francis Galton, who believed in improving human genetic traits through selective breeding. American eugenicists, backed by funding from foundations like the Carnegie Institution and Rockefeller Foundation, promoted the idea of genetic superiority of certain races, particularly Nordic, Germanic, and Anglo-Saxon peoples. They advocated for strict immigration laws, anti-miscegenation laws, and the forced sterilization of the poor, disabled, and "immoral" individuals [54].

Key figures in the American eugenics movement included Charles B. Davenport, Henry H. Goddard, Harry H. Laughlin, and Madison Grant, who lobbied for various eugenic solutions such as immigration restriction, sterilization, segregation, and even extermination. Organizations like the Eugenics Record Office and the American Eugenics Society collected family pedigrees and trained field workers to analyze individuals in mental hospitals and orphanages.

By 1910, a network of scientists, reformers, and professionals actively promoted eugenic legislation and projects in the United States. The American Breeder's Association established a specific eugenics committee, and organizations like the American Association for the Study and Prevention of Infant Mortality advocated for government intervention to promote the health of future citizens.

Some feminist reformers, including Margaret Sanger, the leader of the American birth control movement, supported eugenic legal reform. Sanger incorporated eugenics language to advance her agenda of preventing unwanted children from being born into disadvantaged lives. She even endorsed sterilization to discourage the reproduction of individuals believed to pass on mental disease or physical defects.

In the Deep South, women's associations played a significant role in supporting eugenic legal reform and implementing eugenic institutions segregated by sex to prevent "feebleminded" individuals from breeding.

Various state legislatures enacted eugenic initiatives, such as marriage laws with eugenic criteria and compulsory sterilization bills. Indiana became the first state to pass sterilization legislation in 1907, followed by Washington, California, and Connecticut. The 1927 Supreme Court case Buck v. Bell upheld the constitutionality of forced sterilization, leading to an increase in sterilization rates across the country.

It is indeed ironic that although many of the people who supported eugenics rejected the concept of biological evolution proposed by Darwin as the "survival of the fittest" through "natural selection," they nevertheless believed in what was known as Social Darwinism, in which the strong increase their wealth and power while the weak see their wealth and power decrease [54]. Herbert Spencer promoted these ideas in his work, based on the ideas of Thomas Malthus and Darwin. Spencer compared society to a living organism and argued that, just as biological organisms evolve through natural selection, society evolves and increases in complexity through analogous processes [55].

There is no doubt that the eugenics movement in America influenced the rise of Nazism in Germany [56]. Americans like Henry Ford was a strong supporter of racial theories and supported the Nazi movement [57] and Thomas J. Watson, founder of IBM, supported Nazism through his German subsidiary, even providing the Nazis with lists of Jews [58].

Social Darwinism declined in popularity as a purportedly scientific concept following the First World War [59], and was largely discredited by the end of World War Two due to its association with Nazism and due to a growing scientific consensus that eugenics and scientific racism were groundless.

Chapter 6

The Immigration to America
of European Scientists Fleeing Persecution

1. Jewish Scientists Flee Persecution

In the period 1930-39 before World War Two, as the wave of anti-Semitism engulfing Europe developed, there was a positive tsunami of Jewish scientists of German, Austrian, Hungarian and other nationalities emigrating from Europe to the US. Their estimated number by 1944 was 133,000, and they contributed enormously to the development of basic sciences in the US, including increases in patents and expansion of scientific networks [60].

Among them was a large proportion of high-level scientists, particularly physicists and chemists, some of whom were helped in various ways by US officials to escape Europe, such as Varian Fry [61] and Hiram Bingham III [62]. While the majority of Jews were denied visas and prevented from entering the US, due largely to anti-Semitism among State Department officials [63], the cream of the crop of the scientists were facilitated. Among them were the physicists, Albert Einstein from Germany, Leo Szilard from Hungary, and many others whose names would become synonymous with the leap in American ability in the crucial area for the future war effort of nuclear physics. Some were not Jewish, such as Enrico Fermi from Italy, who emigrated because his wife was Jewish.

Many of these physicists were familiar with the developments being made in nuclear physics in Europe during the period 1900-1930. They knew of the work of Ernest Rutherford in England on the splitting of the atom, of Hans Bethe and Lise Meitner in Germany on nuclear fission and the energy produced when splitting the atom [64], and of Neils Bohr in Denmark on the structure of the atom, and of his German student Werner Heisenberg, who enunciated the famous "uncertainty principle" and who was later to become the Director of the German nuclear program during World War Two. Each of these individuals contributed significantly to the knowledge and understanding of the atom and of its potential to produce enormous amounts of energy.

However, this culture of scientific achievement in nuclear physics was not present in the US. In fact, the most famous American physicist, Robert Millikan, who had won the Nobel prize in 1923 for measurement of the electron, was quoted as saying in 1929, "There is no likelihood to me that man can ever tap the power of the atom, there is no appreciable energy available to man through atomic disintegration" [65]. Ernest Rutherford himself was also skeptical that splitting the atom would result in large amounts of available energy.

Following their arrival in the US, several German Jewish emigres played very important roles in atomic research in America. Einstein was accommodated at Princeton, where he played a role in the Institute for Advanced Studies in expanding knowledge of atomic theory. Fermi went to the University of Chicago, where he famously built the first nuclear reactor core Pile-1 under the stadium of the Chicago University and Szilard worked with Fermi.

2. Szilard and Einstein Author Letter on A-Bomb

Szilard authored the famous letter which Einstein sent under his signature to President Roosevelt on August 2, 1939, warning him of the possibility of the development of an atomic bomb with enormous potential (Figures 10, 11) [66]. This led eventually to the establishment of the Manhattan Project in New Mexico, which was under the scientific direction of Robert Oppenheimer, an American-born Jew.

It is well-known that they did indeed develop the atomic fission bomb and contrary to the original intentions of some of the scientists, two were dropped on the cities of Hiroshima and Nagasaki in order to force the surrender of the Japanese without needing to carry out an invasion of the Japanese Home islands.

What would have happened if these Jewish scientific immigrants had not arrived in the US before World War Two, had they not pursued their research on the atom and had not directly persuaded President Roosevelt to initiate a major and huge commitment to develop atomic fission that resulted in the Manhattan project that led to the Atomic Bomb? There would never have been the A-bombs that were dropped on Hiroshima and Nagasaki by order of President Truman and the War would *not* have ended in August 1945 (VJ Day was Aug 15, 1945), but the US would have had to mount an invasion of Japan itself and there would have been an estimated 1 million US casualties [67].

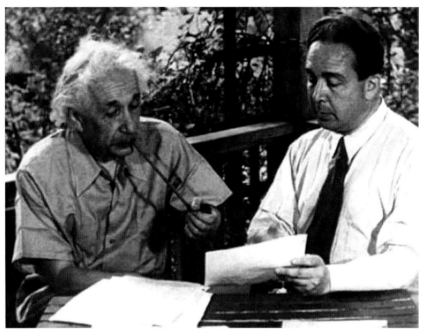

Source: Public domain, Office of Legacy Management, US Dept. of Energy.

Figure 10. Albert Einstein and Leo Szilard 1939.

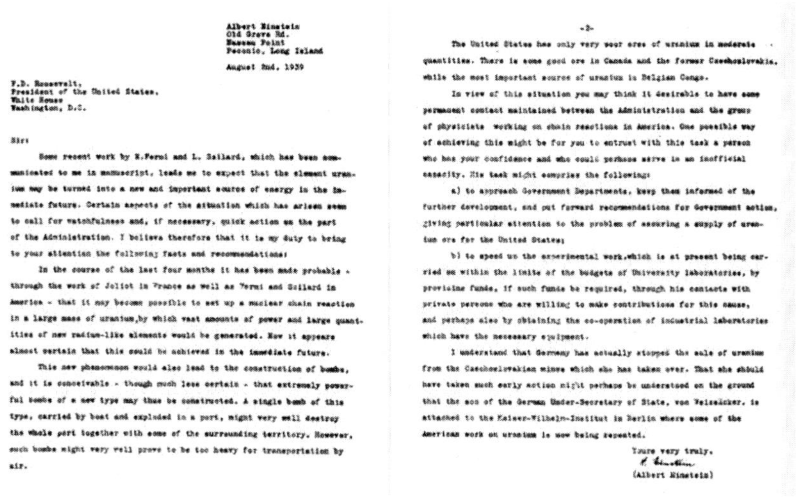

Albert Einstein
Old Grove Rd.
Nassau Point
Peconic, Long Island

August 2nd, 1939

F.D. Roosevelt,
President of the United States,
White House
Washington, D.C.

Sir:

Some recent work by E.Fermi and L. Szilard, which has been communicated to me in manuscript, leads me to expect that the element uranium may be turned into a new and important source of energy in the immediate future. Certain aspects of the situation which has arisen seem to call for watchfulness and, if necessary, quick action on the part of the Administration. I believe therefore that it is my duty to bring to your attention the following facts and recommendations:

In the course of the last four months it has been made probable - through the work of Joliot in France as well as Fermi and Szilard in America - that it may become possible to set up a nuclear chain reaction in a large mass of uranium,by which vast amounts of power and large quantities of new radium-like elements would be generated. Now it appears almost certain that this could be achieved in the immediate future.

This new phenomenon would also lead to the construction of bombs, and it is conceivable - though much less certain - that extremely powerful bombs of a new type may thus be constructed. A single bomb of this type, carried by boat and exploded in a port, might very well destroy the whole port together with some of the surrounding territory. However, such bombs might very well prove to be too heavy for transportation by air.

-2-

The United States has only very poor ores of uranium in moderate quantities. There is some good ore in Canada and the former Czechoslovakia, while the most important source of uranium is Belgian Congo.

In view of this situation you may think it desirable to have some permanent contact maintained between the Administration and the group of physicists working on chain reactions in America. One possible way of achieving this might be for you to entrust with this task a person who has your confidence and who could perhaps serve in an inofficial capacity. His task might comprise the following:

a) to approach Government Departments, keep them informed of the further development, and put forward recommendations for Government action, giving particular attention to the problem of securing a supply of uranium ore for the United States;

b) to speed up the experimental work,which is at present being carried on within the limits of the budgets of University laboratories, by providing funds, if such funds be required, through his contacts with private persons who are willing to make contributions for this cause, and perhaps also by obtaining the co-operation of industrial laboratories which have the necessary equipment.

I understand that Germany has actually stopped the sale of uranium from the Czechoslovakian mines which she has taken over. That she should have taken such early action might perhaps be understood on the ground that the son of the German Under-Secretary of State, von Weizsäcker, is attached to the Kaiser-Wilhelm-Institut in Berlin where some of the American work on uranium is now being repeated.

Yours very truly,
A. Einstein
(Albert Einstein)

Source: Public domain, Franklin D. Roosevelt Library, Hyde Park NY.

Figure 11. Facsimile of the letter send by Einstein to President Roosevelt.

It is not generally known that an attempt at a coup against the Emperor was attempted by elements of the Japanese Army in order to prevent him broadcasting his message of surrender to the Japanese people [68]. Although some 120,000 people were killed by the bombing of Hiroshima and another 65,000 in Nagasaki, given the amount of resistance encountered in the invasion of Okinawa, and the suicides carried out by large numbers of Japanese, particularly women, it can be estimated that there would have been millions of Japanese casualties resulting from an invasion of the Home islands. So ironically in effect the dropping of the Atomic bombs saved lives, both American and Japanese, although there is some controversy about whether or not dropping the second bomb on Nagasaki was indeed necessary.

Chapter 7

Science and Technology Help America Win World War Two

1. Vannevar Bush

If any one man could be regarded as instrumental as the initiator and proponent of support for basic research in the USA, that man would be Vannevar Bush (Figure 12) [9].

He was born in Everett, Massachusetts, in 1890 and went to Tufts University and MIT. Bush played a role in many engineering developments in circuit design and radio technology that led to the development of the Raytheon Company in 1922 that became a large electronics company and defense contractor.

Source: Public domain, Library of Congress.

Figure 12. Vannevar Bush seated at his desk.

At MIT in 1932 he became Vice-President and Dean of Engineering. In 1938 he was appointed President of the Carnegie Institute of Washington, which brought him in close contact with the Government of the USA. He was an engineer, inventor and science administrator, who from its inception in 1941 and during World War Two was Director of the US Office of Scientific Research and Development, and was the first Science Adviser to a US President, President Roosevelt. Although many scientists made contributions towards the development of scientific research in the USA, Vannevar Bush was pre-eminent among them [69].

In 1940, prior to the US joining the War, the British revealed to the US that they had made significant strides in developing radar to detect approaching German airplanes. Realizing the significance of this technology Bush arranged for MIT to develop airborne radar that was available by 1941.

Bush's developments in circuit design had enabled him to effectively develop an analog computer. In the 1930's he developed the 'differential analyzer,' but he also made significant contributions to the development of computing through his concept of the "memex" machine. The memex was essentially a prototype for a hypertext system, and it was described in Bush's famous 1945 essay "As We May Think." In the essay, Bush envisioned a machine that could store and retrieve information in a manner similar to the way the human brain works, with the ability to link and cross-reference information in a nonlinear way. This idea influenced the development of hypertext, which is the underlying technology of the World Wide Web [70].

In 1940, Norbert Weiner approached Bush with a proposal to develop an electronic computer. Bush declined to provide funding because he thought it could not be completed before the end of the war. In this he was correct, but nevertheless, Weiner approached the Army and they provided funding to build what would be known as ENIAC, the world's first electronic computer. Bush was considered short-sighted by many, he refused to provide support for social sciences and also refused to support the development of rockets or missiles. For this he was later criticized.

One of the first applications of science to military technology that Bush oversaw was the proximity fuse that was developed by Merle Tuve and James Van Allen. This was designed to ensure that bombs would explode even if they did not directly hit their target, they only had to be in the proximity of their target. This was not only advantageous because it increased the likelihood of an effective explosion, but also the damage caused by blast was also very significant. When these proximity fuses were used in American ordnance in the first involvement of American forces in North Africa in 1942,

Sir Solly Zuckerman, who was to become the British equivalent to Bush, who was an expert in the effects of bombing, discovered that the American bombs were more efficient at taking out German emplacements than the British ones. When he discovered why, he immediately recommended that the British adopt a similar proximity fuse [71].

2. The Decision to Develop an Atomic Bomb

Perhaps Bush's most significant initiative was his role in persuading the US Government to undertake a program to create an atomic bomb that would become the Manhattan Project. Bush met with President Roosevelt in 1941, and following the initiative of the German Jewish émigré nuclear scientists Szilard and Einstein in 1939 (see Chapter 6) and the British program in atomic development, Roosevelt gave his go-ahead for a crash program to attempt to develop an atomic bomb. The Manhattan Project was funded through Bush's office, but was to be run by the US army under the direction of Secretary of War Henry Stimson and by Brigadier General Leslie Groves and under the scientific direction of Robert Oppenheimer (see Chapter 8).

3. Radar

As the use of aerial attacks revolutionized warfare in World War Two in favor of the attacker, so the use of *radar* reversed the advantage in favor of the defender. With the use of radar you could see the planes coming and thus prepare to counter-attack. That is what happened at the Battle of Britain in 1940, where radar stations set up along the coast of southern Britain and out into the sea enabled the Royal Air Force to defend the United Kingdom against large-scale attacks by Nazi Germany's air force, the Luftwaffe. It was the first major military campaign in history fought entirely by air forces and indeed the first defeat of Germany in World War Two. Radar turned the tables on the attacker.

The name radar comes from *ra*dio *d*etection and *ra*nging, and its origin can be traced back to Heinrich Hertz's experiments in the late nineteenth century, which revealed that metallic objects could reflect radio waves. This concept was initially proposed by James Clerk Maxwell in his pioneering work on electromagnetism. However, it was not until the early twentieth century that practical systems utilizing this principle were developed. A German inventor Christian Hülsmeyer was the first to employ these principles in creating a rudimentary ship detection device in 1904 intended to help prevent collisions in fog.

Robert Watson Watt and Arnold F. 'Skip' Wilkins developed the first radar system in Britain. The system used directional radio signals and oscilloscopes to detect the position of aircraft. The technology was developed in response to the threat of air raids and invasion, and the first successful test was carried out in 1935. The technology was later expanded into the Chain Home system, which used multiple radar stations to detect approaching enemy aircraft during World War Two [72].

Ironically, although radar was largely invented by Germans it helped to defeat Nazi Germany, and it was first applied as a military application by the British. As mentioned above, it was Vannevar Bush who foresaw the need to install radar systems in planes during World War Two. The development of radar was a major advance that helped the Allies win World War Two.

Chapter 8

The Atomic Age

1. Basic Research on Nuclear Physics in Europe

The development of the atomic bomb is a story told in many books [73-75], a documentary [76] and a hit movie entitled "Oppenheimer." The particular aspect that I will emphasize here is the amazing fact that all the basic research in atomic physics that led to the development of the atomic bomb was done in Europe and the applied research that resulted in the actual bombs that were dropped on Japan in 1945 was done in the USA. It is this clear distinction between basic research in Europe and applied research in the US that is the basic theme of this book, and the atomic bomb is an excellent example of this distinction.

Henri Bequerel discovered radioactivity in 1894 at the Museum of Natural History in Paris while investigating uranium emanations. His student Madame Curie (Marie Salomea Skłodowska-Curie) and her husband Pierre Curie in 1896 investigated the natural ore pitchblende containing uranium and discovered the elements polonium (named after her home country) and radium. For this the three of them won the Nobel Prize in Physics in 1903. She also won the Nobel Prize in Chemistry in 1911, and was the first woman to win a Nobel Prize and the first person to win two prizes in different subjects. Subsequently these discoveries were considered to be the basis for therapy of cancer.

The understanding of the basis for subatomic particles came largely from work carried out in Britain. J. J. Thomson discovered the electron in his experiments on cathode rays starting in 1897 in the Cavendish Laboratory in Cambridge, England. For this he was awarded the Nobel Prize in Physics in 1906. The proton was discovered by Thomson's student Ernest Rutherford in 1909, and the neutron was discovered by James Chadwick in 1932.

The first realistic concept of the structure of the atom was proposed by Niels Bohr in 1913, a Danish student of J. J. Thomson. Bohr's model was developed from the proposal of Rutherford that the atom contained a nucleus consisting of protons with electrons circulating around the nucleus. Although Bohr had no explanation why the negatively charged electrons did not fall into the positively charged nucleus, nevertheless his concept of electron orbitals led directly in the quantum mechanical concept of atomic structure that is accepted today.

Nuclear fission was discovered by Otto Hahn, Fritz Strassman and Lise Meitner in the Kaiser Wilhelm Institute for Chemistry in Berlin in 1938 by bombarding uranium with neutrons. Hahn was the only one to receive the Nobel Prize in Physics in 1944 for this work. Lis Meitner was Jewish, and finding her situation in Berlin untenable, she was helped to escape from Germany and moved to Sweden, where she eventually became a citizen and worked at the Institute of Technology in Stockholm. She was nominated numerous times for a Nobel Prize, but later release of documents showed that her Swedish colleague Manne Siegbahn blocked her nomination [77].

It was Meitner who calculated the energy that is produced in nuclear fission and was the first to realize the potential of the tremendous amount of energy that nuclear fission of uranium-235 produces. It was only a brief step from this to realize the concept of a bomb. The first use of the term "atomic bomb" appeared in a letter of American scientist John R. Dunning published in the Physical Review in 1940 [78]. Dr. Dunning, along with colleagues at Columbia University, was conducting experiments on the fission of uranium, and they used the term "atomic bomb" to describe the immense energy released by the process.

It was the culture of basic research and the structure of the organizations to fund it that led to the predominance of European countries in the significant findings in physics and chemistry that characterized the early twentieth century.

2. The Manhattan Project

I have previously described how the project for the development of the atomic bomb was initiated by a letter from Jewish immigrant scientists Albert Einstein and Leo Szilard to President Roosevelt (see Chapter 6). Roosevelt consulted with his scientific adviser Vannevar Bush, who then funded the program for the building of the atom bomb project through the Office of Scientific Research and Development (see Chapter 7). The program was called the Manhattan Project. It was run by the Department of Defense under the direction of General Leslie Grove with J. Robert Oppenheimer as the Scientific Director (Figure 13) [79].

Source: Public domain, UK Government, 1946.

Figure 13. J. Robert Oppenheimer [79].

These two very different men nevertheless managed to work together on this incredible task that was a feat of applied research that could perhaps only have been accomplished in the USA [80].

Oppenheimer was born in New York City to Jewish immigrants from Germany, He earned a bachelor's degree in chemistry from Harvard University in 1925 and a PhD in physics from the University of Göttingen in Germany in 1927. He joined the physics department at the University of California, Berkeley, where he became a full professor in 1936 [79].

He was an unlikely candidate for this job, he was admired by many for his intellectual capability. But he had never organized any major project and had no experience in administration. He was an enigmatic character, both charming and yet aloof, brilliant and yet socially inept. Nevertheless Groves chose him out of all the available candidates because he felt he was a dedicated scientist who could recruit and lead a group of physicists in such a huge and important task.

Oppenheimer had crucially carried out calculations that estimated that for a uranium-235 bomb to achieve criticality would require 2.5-5 kg. If we are to believe the evidence of Werner Heisenberg, who was in charge of the German nuclear program, the Germans made mistakes in their calculations and thought that it would require much more U-235, up to a ton. This was portrayed in the historical play "Copenhagen," by Michael Frayn, dramatizing the meeting between Niels Bohr and Heisenberg that took place in 1941 [81].

The issue of whether or not Heisenberg intentionally prevented Germany from developing an atomic bomb is a matter of conjecture. However, what is known is that Moe Berg, the famous Jewish American baseball player, who spoke fluent German, acted as an American spy during World War Two [82]. He was dispatched to a conference in Zurich in 1944 where Heisenberg was to speak, and it was his mission to determine whether or not Heisenberg was telling the truth when he stated that the German program to develop the A-bomb was stalled. If he was not convinced he was to shoot Heisenberg.

Berg approached Heisenberg after his talk and asked to walk with him back to his hotel in order to ask him a question. Heisenberg agreed, and Berg asked about the A-bomb research project in Germany. Heisenberg later said he felt that he was in danger if he did not answer honestly, so he did. Berg was convinced that Heisenberg was speaking the truth and so, although he had a loaded gun in his pocket, he did not use it.

Source: Public domain, Manhattan Project, US Dept. of Energy.

Figure 14. The atomic bomb dropped on Nagasaki 1945.

It was Oppenheimer who decided to locate the project in the remote New Mexico desert, which he was familiar with from time he had spent there in his youth. It required the Defense Department to build a small city in Los Alamos and through the next few years, Oppenheimer not only recruited most of the best physicists in America, including many immigrant scientists, but also turned out to be a great manager of the project, partly because he was able to keep so much information in his head at one time. There were up to 120,000 people altogether involved in this mammoth project

At first the plan was to shoot a bullet of enriched U-235 or plutonium into a target of the same substance to obtain a critical mass that could produce a chain reaction, with each fission of an atom producing 2 neutrons. But this did not work, so after a year they revised their design to use a ball of enriched U-235 or plutonium with a sphere around it of high explosives that when detonated would *implode* onto the ball and hence cause the critical mass and the chain reaction, resulting in the explosion. Incidentally the uranium was enriched in U-235 at the facility at Oak Ridge Tennessee. Because of the Einstein equation $E = mc^2$ (where E is energy, m is mass and c is the speed of light), it was understood that a small amount of mass should release a huge amount of energy.

Finally in 1945 they were ready to carry out a test on a prototype bomb, and even until the last minute on July 16 at the Trinity site no one knew if it would work, and they were even afraid that they might ignite the whole atmosphere and cause a catastrophe. However the test proved more successful than anticipated, and they were then ready to build a bomb to be used in the war. When the bomb exploded and produced its mushroom cloud Oppenheimer had the presence of mind to quote the Hindu classic the Bhagavad Gita "I am become death, the destroyer of worlds" [79].

It is an astounding fact that it took only 6 years from the Einstein -Szilard letter reaching President Roosevelt's desk until the A-bomb was successfully dropped on Hiroshima and Nagasaki, Japan (Figure 14). Although the US did none of the *basic* research that underlay the understanding of the nuclear processes involved, nevertheless the US was able to carry out the *applied* research and development program of the huge Manhattan Project successfully in such a short period of time. Such is the organizational genius of America.

Originally Oppenheimer and most of the scientific staff envisaged using the bomb on Nazi Germany, because they knew the Germans had a similar atomic bomb project headed by Werner Heisenberg. It was therefore very important that they worked fast to develop the bomb before the Germans. But, in the event, by the time they had the bomb, Germany was defeated and Hitler committed suicide. Also President Roosevelt died in 1945. So then it was up to his successor President Harry S. Truman to make the decision whether or not to use the bomb against Japan.

As mentioned elsewhere, despite many discussions of the moral aspects of the decision to kill hundreds of thousands of people and level cities, nevertheless the decision taken by Truman, and not by Groves or Oppenheimer, was to use the bomb to in fact save lives. It was estimated that at least a million US servicemen's lives would be saved in preventing a land invasion of the Japanese home islands, as well as several million Japanese lives.

Although Oppenheimer later had misgivings about dropping the atomic bombs on Hiroshima (a uranium bomb with 80% enriched U-235) and on Nagasaki (a plutonium bomb), he nevertheless never expressed any regret. He was under the impression that turned out to be false, that such a weapon with such a terrible outcome of destruction would in fact end future wars. But, worse was yet to come.

3. The Hydrogen Bomb

Edward Teller was a Jewish emigre physicist from Hungary, where he had experienced the terrible murderous anti-Semitic hatred of the Hungarian people. At first the dictator of Hungary, Admiral Horthy, was reluctant to adopt the extreme Nazi measures of concentrating and murdering Jews, partly because he realized that Germany, with which he was allied, was likely to lose the war. So in March, 1944, the Nazis occupied Hungary and installed the Iron Cross fascist regime. At this point the atrocities began against the last large Jewish community in Europe.

Edward Teller escaped Hungary and reached the US and was then recruited to work on the Manhattan Project. While working on the development of the fission bomb with Oppenheimer, Teller had his sights on an even more destructive nuclear weapon, the hydrogen or thermonuclear bomb.

The hydrogen bomb is based on the nuclear process that occurs predominantly on the sun and other stars, namely the fusion of two hydrogen atoms into a deuterium atom. Calculations showed that the amount of energy that could be obtained from such a process could be roughly 1,000 times greater than from a fission process. After the successful development and deployment of the atomic fission bomb, Teller and the US Defense Department set their sights on developing a hydrogen bomb in the US [83].

After leaving the Manhattan Project Oppenheimer became Director of the Institute for Advanced Studies at Princeton, where he was in effect Albert Einstein's boss. At meetings of the Committee of the US Atomic Energy Commission tasked with deciding on future activities after the atomic bomb project, Oppenheimer and other members of the Committee voted against continuing to develop the hydrogen bomb. Oppenheimer considered this was "over-kill" because he hoped that the first use of an atomic bomb would also be its last, and that war between countries, because of the tremendous destructive power of the atomic bomb, would end [79]. But he was wrong, wars continued and the Soviet Union soon in 1949 also tested an atomic bomb.

This reluctance to continue with hydrogen bomb research and development was later used against Oppenheimer when his security clearance was later denied in a 1954 hearing of the Atomic Energy Commission. As Director of the Manhattan project it is obvious that Oppenheimer would be subjected to security surveillance. But he seems not to have taken this possibility seriously. Apart from the fact that his brother Robert was a member of the Communist Party and his second wife was also a member, he chose to visit his lover Jean Tatlock in San Francisco, who was also a member of the Party. He spent nights with her and the FBI were concerned that he could be using her as a courier to send top secret information to the Soviets.

The problem was that when confronted with this affair, Oppenheimer was evasive and prevaricated, perhaps out of embarrassment. But, this was taken as a sign of guilt by the authorities. Also Lewis Strauss, who was Chairman of the Atomic Energy Commission, was hawkish and was antagonistic to Oppenheimer, and voted against him. So famously his security clearance was denied in 1954 and he subsequently faded somewhat into obscurity [84].

Oppenheimer still retained his academic position at the Institute for Advanced Studies in Princeton. He died in 1967, but in 2022 the Biden Administration revoked the decision of the AEC to deny his security clearance. It was another case of the communist era "guilt by association" used so effectively by Senator McCarthy. There is no evidence that Robert Oppenheimer ever passed information to any Communist Party member or in any way compromised the security of the USA

Nevertheless the hydrogen bomb was pursued successfully after the decision by President Truman to proceed in 1950 and was a triumph for Edward Teller. Although its terrible destructive power is so much greater than that of the atomic fission bomb, it has never been used in actual warfare.

4. The Soviet Spies

The US authorities were right in being concerned about Soviet espionage, even though the anti-Communist hysteria in the US caused significant suffering to many innocent people. The trial and execution of Julius and Ethel Rosenberg for spying for the Soviet Union in 1953 was surrounded by some anti-Semitic hysteria and probably some falsified evidence.

Nevertheless, there were real cases of spying, from the Cambridge group of Kim Philby, Donald Maclean, Guy Burgess and Anthony Blunt [85], to the actual atomic bomb spies of Klaus Fuchs [86] and Maurice Wilkins, who both worked at Los Alamos. Wilkins was so disgusted by the dropping of the bombs on Japan that he left Los Alamos and returned to England and decided he would only apply his physics to biology in future. He shared the Nobel Prize for the structure of DNA with Watson and Crick for discovering how to make DNA fibers and obtaining high resolution X-ray diffraction patterns from them.

Also, Theodore Hall, who was the youngest physicist recruited to Los Alamos at the age of 18, subsequently provided secret information to the Soviets. Hall worked in the section headed by Klaus Fuchs on developing the detonator for the bomb. Neither of them knew the other was a Soviet spy. This enabled the Soviets to ensure that the information they received was accurate. Out of a misplaced sense of humanism, not knowing the terrible atrocities that Stalin inflicted as a dictator on his own people, they passed secret information to the Soviet Union that enabling Stalin to also develop the atomic bomb more rapidly.

Chapter 9

The American Revolution in Science after World War Two

1. The Founding of the National Science Foundation

After World War Two, the Office of Scientific Research and Development was no longer needed and was disbanded. However, Vannevar Bush realized that there was a great need for a peacetime agency to replace the function of OSRD in promoting science and technology for the national interest. He wrote an essay in 1945 that is considered one of the most influential papers relating to science and technology ever produced in the USA.

It was entitled "Science, The Endless Frontier," that was a Report to President Roosevelt [87], urging the establishment and funding of a national peacetime research organization. Note that Vannevar Bush's Report to Pres. Roosevelt came 36 years after the similar memorandum by Adolf von Harnack to Kaiser Wilhelm II; a measure of the lag in US appreciation for the significance and impact of basic scientific research on society.

The change from President Roosevelt, who died in 1945, to President Truman, resulted in a significant loss of influence for Vannevar Bush. Through many political changes and compromises over a period of 5 years, Bush's Report finally resulted in Congress establishing the National Science Foundation (NSF) in 1950 to fund basic research in the USA.

One other significant influence that Vannevar Bush exerted on science in the US after World War Two was his concept of large-scale data manipulation needed for the pursuit of science, something that he called Memex. He had thought about this since the 1930's and he crystallized his ideas in an article entitled "As We May Think," that was published in *The Atlantic* magazine in 1945. It consisted of a data storage device in the form of microfilm that could be rapidly switched to enable rapid access to different information. This paper was a forerunner of what we call "the information age," and was extremely influential in the thinking of people who set about using electronic means to develop the mouse, the computer and the internet [88].

2. The Long-term Effect of Immigration of Jewish Scientists

Apart from their singular influence on the developments in nuclear physics which resulted in the atomic bomb, European Jewish immigrant scientists had a widespread salutary effect on American science. This is attested to by the general increase in scientifically based patents produced in subsequent years following their immigration and the development of a much wider range of research on basic scientific subjects [60].

Note that Jews have been awarded 26% of Nobel Prizes in physics, 20% in chemistry, 27% in physiology or medicine and 41% in economics. Although Jews are only 0.25% of the world's population, they have received a staggering 24% of all Nobel Prizes in science (physics, chemistry and physiology or medicine). During the period from 1901-1939 people of Jewish ancestry won 15% of German Nobel prizes while being less than 1% of the German population. From 1939 onwards, when there were no longer any Jews in Germany, the number of Nobels won by Germans did not increase significantly for some time (Figure 1, in Chapter 2), but this is not surprising since Germany had lost the War and was devastated. Similarly, after World War Two the increase in number of Nobels in the UK, which had won the War but was similarly devastated, grew only very slowly. But the US experienced a sharp increase in Nobels following World War Two and surpassed the individual European nations after 1960 (Figure 1), as both its population and expenditure on Research and Development significantly increased.

Countries with increased Research and Development expenditures demonstrate higher growth performance with higher levels of GDP per capita than other countries [89-91]. This salient fact indicates that apart from the influence of the European émigré scientists and the subsequent influence after World War Two of German scientists transferred to the US, particularly in the area of rockets and aeronautics, it was the decision of the US to expend a large amount of funding on research and development after the War that led to its accumulating wealth in that period.

3. Operation Paperclip

After World War Two, it was perhaps a shock to the Western allies to find that the Germans had been so far ahead in various areas relating to military technology. For example, in the development of rockets, such as the V2, that could be fired into the stratosphere and then crash into a city far away and cause enormous damage. The Allies had no such weapons. Also, the Germans developed the first functioning jet airplane, the Me-262 (called the Schwalbe or swallow), that was used in combat at the end of the War. It could easily outfly the propeller planes of the Allies, although it was developed in 1942, not enough of them were produced to affect the outcome of the War.

To obtain the secrets of German research on these and other technologies, there was a race between the US and the USSR as the War came to an end to capture and use the expertise of the German scientists. Those who were caught and transferred to the US were, of course, quite happy not to share the fate of the rest of the Nazi apparatus they had served. This US operation was called Operation Paperclip and resulted in ca. 1,600 German scientists and engineers being transferred to the US [92].

Werner von Braun, the Head of the German rocket program under the Nazis, whose products, including the V2, that killed tens of thousands of slave laborers in their construction and many Londoners as their targets, was never charged with any war crime. Instead, he was appointed Head of the US Army's ballistic missile program and then Head of The National Aeronautics and Space Administration (NASA) space program. The reason was, of course, to try to beat the USSR in the development of rockets and in space exploration. He received the US National Medal of Science in 1975.

Chapter 10

America Becomes the World Leader in Science and Technology

1. Federal Research Spending

The tremendous increases in funding for basic scientific research in the US after World War Two had a profound impact on the world. The National Institutes of Health (NIH) and the National Science Foundation (NSF) played key roles in supporting research in the health-related medical sciences and the basic sciences, respectively. These investments resulted in numerous breakthroughs in our understanding of the biological and physical world, and in many cases, led to revolutionary inventions and applications.

Louis Pasteur's belief that science is indivisible is evident in the close relationship between basic and clinical research at the NIH. The goal of this collaboration is to develop better strategies for treating complex diseases. Clinical investigators work alongside basic science colleagues on the NIH campus to achieve this aim. In addition, the NIH organizes Consensus Development Conferences, which bring together experts from around the world to evaluate existing therapies and assess new modes of therapy. The first such conference was held in 1977 and recommended mammography as a routine diagnostic tool for breast cancer in women over fifty. Since then, over 100 conferences have played a crucial role in disseminating research findings on medical devices, drugs, and surgical procedures to physicians in practice [19].

The discoveries made by NIH -supported investigators are too numerous to enumerate comprehensively. Over 80 Nobel prizes have been awarded for NIH-supported research, with five of them going to investigators in the NIH intramural programs. These discoveries include deciphering the genetic code that governs all life processes, explaining how chemicals transmit electrical signals between nerve cells, and understanding how the chemical composition of proteins affects their biological activity. These fundamental research findings have led to improved comprehension of genetically based diseases, better antidepressants, and drugs that specifically target proteins involved in particular disease processes. Long-term research has also challenged preconceived notions that morbidity and dementia are normal parts of the aging process. Furthermore, certain cancers have been cured and the number of deaths resulting from heart attack and stroke have significantly decreased due to basic and applied research. Research has also shown that preventive measures such as a balanced diet, regular exercise, and not smoking can reduce the need for therapeutic interventions and save healthcare costs.

Dr. Joseph Kinyoun, the founder of the NIH in 1887, likely couldn't have envisioned the NIH's current size and breadth of funding and research programs. The NIH's rich history and scientific opportunities have positioned it as a premier medical research institution that will continue to make significant contributions to human health in the twenty-first century.

The NSF went through a parallel increase in significant funding and research output. After the 1957 Soviet Union orbited Sputnik 1, the first ever man-made satellite, national self-appraisal questioned American education, scientific, technical and industrial strength and Congress increased the NSF appropriation for 1958 to $40 million [93].

Over the next decades Congress would markedly increase funding of the NSF and the NIH. Today they receive billions of dollars, and various institutes and centers within the NIH were created for specific disease-related programs. The NIH's functions were divided into two, the intramural research program, and the extramural grant program. Each Institute has its own separate intramural and extramural programs designed to advance knowledge and understanding of disease and therapy in each of the major disease categories and to support research through competitive grants at Universities and Medical Schools throughout the USA and the world.

Source: Reproduced with permission, University of Firenze Press.

Figure 15. US Federal Government spending on research (in billions of 2017 dollars) [94].

Between them, NIH and NSF funding account for most of the biomedical and scientific research carried out in the USA, and constitute the largest commitment of any country around the world to the funding of basic research (Figure 15) [94]. The significant increase in scientific research in the US after World War Two roughly parallels the increase in the number of Nobel Prizes won (Figure 1). Furthermore, long after World War Two, when the two countries Britain and Germany had recovered their former economic strength, they never did come anywhere close to the scientific standard that the USA had attained.

2. The Laser

The word laser comes from "light amplification by stimulated emission of radiation." It consists of a coherent beam of light that can be focused on a single spot. Theodore Maiman operated the first laser on 16 May 1960 at the Hughes Research Laboratory in California, by shining a high-power flash lamp on a ruby rod with silver-coated surfaces. He promptly submitted a short report of the work to the journal Physical Review Letters, but the editors turned it down. However, it was later recognized as having been significant.

It was soon realized that the laser was far more important than the maser which preceded it based on microwave radiation. The laser led to such common applications as neon signs, as well as automated cutting devices, bar codes and laser printers [95]. They have also had important applications in medicine, especially in surgery.

Perhaps the most significant application of lasers has been in long-distance communication through fiber-optic cables using multiplexing, enabling digital information to be transferred across oceans in seconds.

High power lasers are also being used in fusion energy production. These ideas were developed in the 1970's [96] and finally brought success in 2023 [97]. Recently a laser weapon called the Iron Beam has been described by the Israeli defense company Rafael [98], which can shoot down missiles at much lower cost than expensive anti-missile missiles like the Iron Dome system. This could have a significant influence on future conflicts.

3. The Development of Electronics

A semiconductor device possessing a minimum of three terminals for connection to an electrical circuit is known as a transistor. Typically, the third terminal governs the flow of current between the other two terminals, making it useful for amplification (as in radio receivers) or for rapid switching (as in digital circuits). The transistor supplanted the vacuum-tube, which was substantially larger and required significantly more power to function.

Bell Laboratories was founded in New York City in 1925 and then moved to Murray Hill, New Jersey, as the research labs of AT & T. On December 23, 1947, Bell Labs successfully demonstrated the first transistor, marking a breakthrough in the field of electronics. William Shockley, John Bardeen, and Walter Brattain are credited with its invention [99].

This tiny device made it possible to amplify and rapidly switch electronic signals, and it revolutionized the electronics industry. The advent of the transistor is widely regarded as one of the most consequential discoveries in human history.

Canadian physicist Julius Edgar Lilienfeld filed the first patent for the transistor principle in 1925, but his device received no attention as he did not publish any articles about it. In 1934, German physicist Dr. Oskar Heil patented another field-effect transistor.

William Shockley and his co-worker Gerald Pearson at Bell Labs successfully built an operational version based on Lilienfeld's patent, but they did not acknowledge Lilienfeld's work in their papers or articles [100].

The Bell Lab's work on the transistor originated from their efforts during World War Two to produce highly pure germanium crystals for diodes used in radar. After the War, Shockley took up the challenge of creating a functional semiconductor device. He secured funding and lab space and collaborated with Bardeen and Brattain to solve the problem.

In 1950, Shockley developed an improved type of solid-state amplifier which was licensed to a number of other electronics companies, including Texas Instruments, who produced a limited run of *transistor radios*. Thus the ubiquitous transistor radio was born and the rest as they say is history. In 1956, John Bardeen, Walter Brattain, and William Shockley were awarded the Nobel Prize in Physics for their work on semiconductors and their discovery of the transistor effect.

The *printed circuit*, which was developed during the War and released for commercial use in 1948, was key to miniaturizing electronics and making them more reliable. Prior to the invention of printed circuit boards (PCBs), electrical and electronic circuits were wired point-to-point on a chassis, typically a metal frame. The components were attached to the chassis using insulators, and their leads were connected directly or with jumper wires by soldering. This made the circuits large, bulky, heavy, and relatively fragile. Moreover, production was labor-intensive, making the products expensive.

The development of modern PCB methods began in 1903, when a German inventor named Albert Hanson described flat foil conductors laminated to an insulating board. Thomas Edison experimented with chemical methods of plating conductors onto linen paper in 1904. Arthur Berry in 1913 patented a print-and-etch method in the UK, and in the United States, Max Schoop obtained a patent for flame-spraying metal onto a board through a patterned mask. Charles Ducas in 1925 patented a method of electroplating circuit patterns. Preceding the invention of PCBs, John Sargrove's 1936–1947 Electronic Circuit Making Equipment (ECME) sprayed metal onto a Bakelite plastic board in a similar manner. The first actual printed circuit was invented by the Austrian engineer Paul Eisler while working in the UK around 1936. He incorporated it into a radio set. In 1941, a multi-layer printed circuit was used in German magnetic naval mines [101].

Around 1943, the US began using PCBs on a large scale to make proximity fuses for use in World War II. These fuses required an electronic circuit that could withstand being fired from a gun and be produced in quantity. The Centralab Division of Globe Union proposed a viable solution: using a ceramic plate that was screen-printed with metallic paint for conductors and carbon material for resistors, with ceramic disc capacitors and subminiature vacuum tubes soldered in place. This technique proved successful, and the resulting patent on the process, which was classified by the US Army, was assigned to Globe Union.

It was not until 1984 that Harry W. Rubinstein was recognized by the Institute of Electrical and Electronics Engineers (IEEE) for his early key contributions to the development of printed components and conductors on a common insulating substrate. PCBs revolutionized the electronics industry as they allowed not only wiring but also passive components to be fabricated on the same substrate. This represented a significant step in the development of integrated circuit technology.

The current technology to produce PCB's is to design the printed circuit in large size and then miniaturize it using a "stepper machine" and micro-photolithography to image the circuit layers and etch the features on a silicon chip. Due to their ubiquity with a silicon base they gave their name to 'Silicon

Valley' in California, where much of the advances in electronics and computers later took place. Silicon chips are today a b$60 industry [102].

4. Computers and the Internet

Perhaps the most significant invention of the twentieth century was the development of the *electronic computer*, which emerged as a result of advances in both hardware and software. There are many books written on this subject [103-105] and many of the details are well known. But I will summarize the subject here.

Alan Turing, a British mathematician is credited with the first true concept of a machine that could both calculate and think. His application of this to the cryptanalysis of the German secret codes of the enigma machine at Bletchley Park during World War Two has been widely dramatized. He developed a statistical computer machine called the *bombe* that was installed in 1940 [106].

The first actual electronic digital computer was the ENIAC (Electronic Numerical Integrator and Computer) built by the US army in 1945, initially to calculate artillery trajectories. According to the Computer History Museum of Mountain View, California [107], the Kenbak-1 is considered by to be the world's first personal computer [108]. It was designed and built by John Blankenbaker of Kenbak Corporation in 1970, and was first sold in 1971. Unlike a modern personal computer, the Kenbak-1 was built with small-scale integrated circuits, and did not use a microprocessor, which only became available in 1971 [109].

The first widely used personal computers were built by Steve Jobs and Steve Wozniak in a garage that led to the founding of the Apple Computer Company in 1976 and the development of the Apple II computer in 1977 and the Mac computer in 1984.

At first it was thought that the hardware was the most important element in the computer, but then software became essential. Microsoft Corporation was founded in 1975 by Bill Gates and Paul Allen and the first disk operating system (DOS) software was developed in a small office at Albuquerque airport. When the huge IBM Company started producing personal desk top computers in 1981 using Microsoft DOS, the future became imminent. These inventions enabled computing power to be available to individuals and businesses in a way that was previously unthinkable.

These developments in computing led to the micro-miniaturization that eventually allowed the smart phone to be developed, whereby today almost every individual in the Western World now carries around in their pocket or handbag. The first handheld smart phone precursor was developed by an IBM engineer Frank Canova in 1992. In 2007, Steve Jobs at Apple Computer introduced the iPhone that was followed by the Android operating system from Samsung in 2017. Smartphones have become essential to life on earth and now we cannot live without them!

The history of the *Internet* traces back to the late 1950s when computer science was emerging as a discipline. Scientists and engineers sought to build interconnected computer networks and explore the concept of time-sharing between computer users. In the United States, J. C. R. Licklider at the Advanced Research Projects Agency (ARPA) of the Department of Defense envisioned a universal network [110].

At the same time, Paul Baran at the RAND Corporation proposed a distributed network, and Donald Davies conceived of packet switching in the United Kingdom. In 1969, ARPA awarded contracts for the development of the ARPANET project, which incorporated the packet switching technology proposed by Baran and Davies. Commercial Internet service providers (ISPs) emerged in the late 1980's, and the ARPANET was decommissioned in 1990.

Tim Berners-Lee's work at CERN in Switzerland in the late 1980s resulted in the World Wide Web (www), revolutionizing communication, commerce, and technology [111]. The Internet's capacity expanded dramatically in the mid-1990s with the introduction of fiber optic cables and this enabled near-instant communication through various online services that we know today.

5. The Development of Silicon Valley

When one thinks of science and technology in America one naturally thinks of Silicon Valley, where so many start-ups have been born that have revolutionized our lives. Silicon Valley is located in California at the southern tip of San Francisco Bay around the city of San Jose. It got its name because of the ubiquity of the element silicon in the transistors and electronic chips that are the basis of much that goes on in the high-tech start-ups.

The area rose to dominance in the electronics field because of a fortunate combination of factors. The presence of a large US Navy base in San Francisco led to them taking over the air field at Sunnyvale near San Jose. This later became a hub for technology, mainly for the Navy in terms of improved communications. The presence of Stanford University, founded in 1891, led to an infusion of technically qualified graduates. Cyril Elwell, a graduate of Stanford University, acquired the U.S. patents for Poulsen arc radio transmission technology (Figure 16) and established the Federal Telegraph Corporation (FTC) in Palo Alto. In the following decade, the FTC pioneered the development of the world's first global radio communication system. Notably, in 1912, they secured a significant contract with the Navy, developing maritime communication [112].

Over time the airfield known as Moffett Field attracted many more companies, the largest being Lockheed Aeronautics. The first IBM plant was established in San Jose in 1943. In the 1950's Stanford University founded an industrial park known as the Stanford Research Park, and this attracted more technologically driven companies to the area. The first tenant there was Varian Associates that had developed radar for the military during the War. Another famous company that moved in was Hewlett-Packard, founded by Stanford graduates Bill Hewlett and David Packard in the latter's garage in 1939 [113].

Another inventor who moved to nearby Mountain View was William Shockley, who had led the group that developed transistors at Bell Labs. In 1956 he established a company called Shockley Semiconductor Laboratories. This attracted a group of experts in the field, who then left and formed their own company Fairchild Semiconductor, and then two of them split off and formed the famous Intel Corporation.

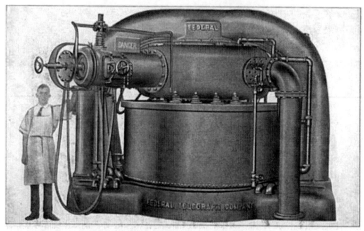

Source: Public domain image from 1919.

Figure 16. The Poulsen arc transmitter invented by Danish engineer Valdemar Poulsen in 1903. It was used in early wireless telegraphy to convert direct current electricity into radio frequencies.

Working at Bell Labs, Mohamed Atalla in 1957 introduced a groundbreaking technique that electrically stabilized silicon surfaces. This pivotal development led to the displacement of germanium with silicon as the dominant material in semiconductor technology and laid the foundation for the large-scale production of silicon semiconductor devices. Building upon this achievement in 1959 Atalla, with his colleague Dawon Kahng, invented the MOS (metal-oxide-silicon field-effect transistor).

The MOS transistor represented a significant leap forward as the first compact transistor capable of miniaturization and mass production for a wide range of applications. This breakthrough is widely acknowledged as the catalyst for the silicon revolution, transforming the semiconductor industry and shaping the modern technological landscape. In 1961 Attala moved to Hewlett-Packard and the MOS technology fueled the development of many electronics based companies in the area, leading to the origin of Silicon Valley [114].

The Homebrew Computer Club was a casual gathering of electronics enthusiasts and technically inclined hobbyists who came together to exchange components, circuits, and knowledge related to do-it-yourself construction of computer devices. The group was founded by Gordon French and Fred Moore. Their shared interest in creating a regular and inclusive platform drove them to establish the club, aiming to make computers more accessible to everyone. The inaugural meeting took place in March 1975 at French's garage located in Menlo Park, California. Steve Wozniak and Steve Jobs attribute their inspiration to the discussions and interactions at this initial gathering. The Homebrew Computer Club played a pivotal role in shaping their vision, ultimately leading to the design of the original Apple I and its successor, the Apple II computers [115].

The TRS-80 Micro Computer was a desktop micro-computer launched in 1977 and sold by Tandy Corporation through their Radio Shack stores. The name is an abbreviation of *Tandy Radio Shack, Z80 [microprocessor]*. It was one of the earliest mass-produced and mass-marketed retail home computers. It had a significant impact on the proliferation of home computers and was a great learning tool for many young budding computer experts [116].

Silicon Valley has become most famous in recent years for innovations in software and Internet services. The companies from this region have significantly influenced computer operating systems, software, and user interfaces [117]. With support from US sources, Douglas Engelbart invented the mouse and hypertext-based tools in the mid-1960s and 1970s while at Stanford Research Institute. Xerox hired some of Engelbart's researchers beginning in the early 1970s. In turn, in the 1970s and 1980s, Xerox's Palo Alto Research Center played a pivotal role in developing programming, graphical user interfaces, Ethernet, and laser printers.

The Internet became practical and grew slowly throughout the early 1990s. In 1995, commercial use of the Internet grew substantially and the initial wave of internet startups such as Amazon.com, and eBay began operations [118].

7. The Jet Engine

A jet engine is a reaction engine that generates propulsion by discharging a rapid, heated gas stream. Though this definition may encompass rockets, water jets, and hybrid propulsion, the term usually refers to an internal combustion air-breathing engine.

The concept of a jet engine is not new. However, it wasn't until the twentieth century that the required technological advancements to realize this idea were achieved. The earliest demonstration of jet power can be traced to the aeolipile, a device documented by Hero of Alexandria in first-century Egypt. The device directed steam power through two nozzles to spin a sphere quickly on its axis. It was seen more as a curiosity, while practical turbine applications can be seen in water wheels and windmills.

The first air-breathing jet engines were hybrid designs, where an external power source compressed the air first. This air was then mixed with fuel and burned to produce jet thrust. Even before the start of World War II, engineers began to recognize that propeller-driven engines were nearing their limits of efficiency. This was the motivation behind the development of the gas turbine engine, the most common form of jet engine.

The gas turbine, a concept used to extract power from the engine to drive the compressor, was not new. The first patent for a stationary turbine was granted to John Barber in England in 1791. The first self-sustaining gas turbine was built by Norwegian engineer Ægidius Elling in 1903. However, these engines did not reach manufacturing stages due to concerns regarding safety, weight, and, most importantly reliability.

The principle of using a gas turbine to power an aircraft was patented in 1921 by Maxime Guillaume. His engine, an axial-flow turbojet, was never built, as it would have required significant advancements in current technology. Alan Arnold Griffith published an "An Aerodynamic Theory of Turbine Design" in 1926,

In 1928, RAF College Cranwell cadet Frank Whittle formally submitted his ideas for a turbojet to his superiors. He submitted his first patent in England on January 16, 1930. The patent showed a two-stage axial compressor feeding a single-sided centrifugal compressor. However, Whittle was unable to interest the government in his invention [119].

In Spain, pilot and engineer Virgilio Leret Ruiz was granted a patent for a jet engine design in 1935. Initial construction began at the Hispano-Suiza aircraft factory in Madrid in 1936, but Leret was executed months later after unsuccessfully defending his seaplane base at the beginning of the Spanish Civil War. His plans, were secretly given to the British embassy in Madrid a few years later by his wife. In Germany, in 1935, Hans von Ohain started working on a design similar to Whittle's. Ohain's first device was experimental and could only run under external power, but he demonstrated the basic concept. Ohain was then introduced to Ernst Heinkel, a prominent aircraft designer, who immediately recognized the potential of the engine. Their subsequent designs culminated in the gasoline-fueled HeS engine which was fitted to Heinkel's compact He 178 airframe and flown in 1939 and was the world's first jet plane.

Meanwhile, in Britain, the Gloster E28/39 had its maiden flight on May 15, 1941, and the Gloster Meteor finally entered service with the RAF in July 1944. These were powered by turbojet engines from Power Jets Ltd., set up by Frank Whittle (Figure 17). The first two operational turbojet aircraft, the Messerschmitt M-262 and then the Gloster Meteor, entered service within three months of each other in 1944.

Following the end of the war, the German jet aircraft and jet engines were extensively studied by the victorious allies and contributed to work on early Soviet and US jet fighters. The legacy of the axial-flow engine is seen in the fact that practically all jet engines on fixed-wing aircraft have had some inspiration from this design.

By the 1950s, the jet engine was almost universal in combat aircraft, with the exception of cargo, liaison and other specialty types. By this point, some of the British designs were already cleared for civilian use, and had appeared on early models like the de Havilland Comet and Avro Canada Jetliner. By the 1960s, all large civilian aircraft were also jet powered, leaving the piston engine in low-cost niche roles such as cargo flights.

Source: Public domain, US National Air and Space Museum.

Figure 17. The Whittle turbojet engine flew in the Gloster Meteor.

8. The History of the Automobile

One cannot talk about the history of science and technology in America without considering that archetype of American culture, the automobile. Although originally perfected by notable European figures like Gottlieb Daimler, Karl Benz, Nicolaus Otto, and Emile Levassor, the automobile's greatest social and economic impact took place in the United States. The 1901 Mercedes (named after his daughter), designed by Wilhelm Maybach for Daimler Motoren Gesellschaft, is considered the first modern motorcar (Figure 18), boasting impressive features and performance for its time [120].

Duryea brothers, bicycle mechanics from Massachusetts, produced the first successful American gasoline automobile in 1893, followed by the first American car race win in 1895. In 1899, thirty American manufacturers produced 2,500 motor vehicles, with an additional 485 companies joining the industry in the subsequent decade.

Source: Public domain image in the USA.

Figure 18. The Mercedes Mark 1.

Source: Public domain, Library of Congress.

Figure 19. Duisenberg 1920's model.

The European design demonstrated its superiority when comparing the 1901 Mercedes with the early American automobiles, like Ransom E. Olds' 1901-1906 one-cylinder, three-horsepower, tiller-steered Oldsmobile. Despite the advanced design of the Mercedes, American ingenuity focused on reconciling top-notch features with affordable pricing and low operating costs, ultimately becoming a triumph in the United States (Figure 19).

In 1908, Henry Ford revolutionized the market with the introduction of the Model T, while William Durant founded General Motors. These events marked a pivotal moment in the automotive industry's history.

The United States, with its vast land area and scattered settlements, had a significantly greater need for automotive transportation compared to European nations. This demand was further fueled by higher per capita income and a more equitable income distribution in the country.

Due to the American manufacturing tradition, it was inevitable that cars would be produced in larger volumes and at lower prices than in Europe. The absence of tariff barriers between states facilitated wide geographic sales. Furthermore, the availability of cheap raw materials and a shortage of skilled labor encouraged early mechanization of industrial processes in the United States.

As a result, products like firearms, sewing machines, bicycles, and other items were standardized and produced in large quantities. In 1913, the United States manufactured about 485,000 of the world's total of 606,124 motor vehicles [121].

The Ford Motor Company, led by Henry Ford, excelled in combining cutting-edge design with affordability. The Ford Model N (1906-1907), a four-cylinder, fifteen-horsepower car priced at $600, was lauded as a low-cost motorcar, garnering overwhelming orders. Ford's improved production equipment enabled them to deliver a hundred cars a day after 1906.

Inspired by the success of the Model N, Henry Ford aimed to create an even better "car for the great multitude." Thus, the four-cylinder, twenty-horsepower Model T was introduced in October 1908, retailing at $825. The Model T boasted a two-speed planetary transmission for easy driving, a detachable cylinder head for simple repairs, and a high chassis to navigate rough rural roads. Vanadium steel was used to make the Model T lighter and sturdier, while innovative casting methods helped keep the price affordable.

With a commitment to large-volume production, Ford revolutionized mass production techniques at the Highland Park, Michigan, plant, which opened in 1910. The moving assembly line was later introduced in 1913-1914. The Model T runabout was sold for $575 in 1912, which was less than the average annual wage in the United States. By the time the Model T was discontinued in 1927, its price had dropped to $290 for the coupe, and an astounding 15 million units had been sold. This mass personal mobility marked a significant milestone in automotive history.

Ford's mass production techniques were quickly adopted by other American automobile manufacturers, while European automakers didn't embrace them until the 1930s. However, this adoption of mass production led to higher capital investments and increased sales volumes, which ended the era of easy entry and free-wheeling competition among numerous small producers in the American automotive industry. Between 1908 and 1929, the number of active automobile manufacturers dropped significantly from 253 to only 44, with Ford, General Motors, and Chrysler (acquired by Walter P. Chrysler in 1925) accounting for around 80 percent of the industry's output. The Great Depression further eliminated most of the remaining independent manufacturers, with companies like Nash, Hudson, Studebaker, and Packard struggling to survive but eventually collapsing in the post-World War Two period.

In Belfast, Northern Ireland, in 1888, a practical pneumatic tire was created by John Boyd Dunlop, a Scot who owned a veterinary practice. This innovation was born out of Dunlop's desire to alleviate the discomfort experienced by his 10-year-old son, Johnnie, while riding his bicycle on uneven pavements. Dr. John Fagan was the family doctor who had recommended cycling as a form of exercise for the young boy. In 1889, cyclist Willie Hume showcased the superiority of Dunlop's tires by achieving victory in the first-ever races featuring these pneumatic tires.

Charles Goodyear invented vulcanized rubber and in 1898 Charles Seiberling founded the Goodyear Tire and Rubber Company in Akron, Ohio. During the 1920s, laboratories at Bayer successfully pioneered the creation of synthetic rubbers. Shortages of rubber in the United Kingdom during WWII spurred research into alternatives for rubber tires. In 1946, Michelin introduced the radial tire construction method. Recognized for its superior handling and fuel efficiency, this innovation rapidly gained traction world-wide [122].

By 1927, automakers were facing a stagnant market as incomes of the day did not support further expansion. To address this situation, installment sales for cars were introduced, eventually becoming a common practice, transforming the purchasing habits of the middle class, and becoming a crucial part of the American economy.

Market saturation coincided with a lack of significant technological advancements. The basic differences between post-World War Two models and the Model T were established by the late 1920s, with only a few additional innovations introduced in the 1930s. General Motors, under the leadership of Alfred P. Sloan, Jr., responded to these challenges by innovating "planned obsolescence" and emphasizing styling. This approach included regular cosmetic changes and major restyling every few years to encourage consumers to trade in their current cars for newer, more expensive models before the cars' useful life had ended.

Sloan's philosophy was focused on making money rather than leading in design or taking risks with unproven experiments. During this period, Ford lost its position as the leading seller of low-priced vehicles to Chevrolet in 1927. By 1936, GM had captured 43 percent of the U.S. market, pushing Ford to third place with 22 percent, while Chrysler held 25 percent. Despite the collapse of automobile sales during the Great Depression, GM managed to consistently earn profits, retaining industry leadership until 1986 when Ford surpassed it in profits.

The automobile industry played a crucial role in producing military vehicles and materials during both World Wars. While civilian vehicle production ceased in 1942 due to the war effort, the war's end saw a significant pent-up demand for new cars. In the postwar era, models and options proliferated, cars became larger, more powerful, and gadget-filled every year, prioritizing profits over practicality. This approach led to decreased engineering focus on economy, safety, and quality. American-made cars were delivered to buyers with numerous defects, many of which were safety-related.

However, the automotive industry experienced significant changes starting in the 1960s and 1970s. Federal standards for safety, emission control, and energy consumption were imposed, and escalating gasoline prices following oil shocks prompted a shift towards more fuel-efficient and well-built smaller cars. Imports, particularly from Japan, started gaining a substantial share of the U.S. market, leading to a decline in American-made car sales. The American automotive industry responded by undergoing massive organizational restructuring and technological advancements. Leaner and more efficient companies were created, focusing on quality manufacturing and employee motivation.

Throughout the twentieth century, the automobile had a transformative impact on American society [123]. It spurred the growth of various industries, influenced recreational and tourism activities, and led to significant infrastructure development, including the construction of streets and highways. It became a driving force behind various ancillary industries, revolutionizing steel, petroleum, and other sectors to meet the demands of the automotive world.

The advent of the automobile marked a significant turning point in American history. It not only put an end to rural isolation by bringing urban amenities, better medical care, and improved schools to rural areas, but also paradoxically rendered traditional family farms obsolete with the introduction of the farm tractor. The automobile, along with trucking, played a pivotal role in shaping the modern city and its surrounding industrial and residential suburbs.

Beyond the physical landscape, the automobile had a profound impact on the architecture of American homes, the structure of urban neighborhoods, and the lives of homemakers. It liberated homemakers from the confines of their homes, granting them newfound mobility and freedom. The automobile became an integral part of American life, and by 1980, nearly 87.2 percent of households owned one or more vehicles, with more than half owning multiple cars. Car sales were predominantly for replacement purposes, showcasing the nation's strong auto-dependence.

However, as the years passed, other forces like electronic media, lasers, computers, and robots emerged, taking precedence in shaping the future. The era of the "Automobile Age" was slowly transitioning into a new "Age of Electronics," where these technologies will define the way Americans work, live, and interact in the future.

9. Travel in the USA

It is well known that the development of the railway system revolutionized travel in the US during the nineteenth century, and allowed fruit from California to be on the tables of residents of the East Coast within a few days. Of course, Britain was ahead of the US in introducing steam trains and rapid travel by rail. It was the decision of President Abraham Lincoln to go ahead with the expansion of the trans-continental railroad system. He realized that by connecting the east and west coasts together with rapid transport this would help to give the USA the concept of being a single country. He signed the Pacific Railway Act of 1862 that chartered the Union Pacific and Central Pacific railroad companies to build this system.

In 1919, Dwight D. Eisenhower, then an officer in the US Army, was part of an expedition that drove across the USA. This experience of the terrible road conditions led him to understand the need for the building of a large interstate highway network. In 1938, President Franklin D. Roosevelt gave Thomas MacDonald, Chief at the Bureau of Public Roads, the go-ahead to start planning a trans-continental highway road system. After World War Two, when he became President, Eisenhower signed the Federal Highway Act of 1956 that instituted the Interstate Highway system, with a final budget of $114 billion. This was by no means an easy task and required, like the railroads, the cutting of tunnels through the Rocky Mountains. But it was accomplished and in 1986 anyone could drive from coast to coast in a few days (Figure 20) [124, 125].

Even more efficient travel resulted with the introduction of commercial flights that reduced the time from coast to coast to a matter of hours. Following the developments of airplanes during World War One it was soon after, as described elsewhere (see Chapter 5), that Trans-World Airways was developed. The jet engine was first invented in England by John Whittle in 1928, but the first operational jet-powered plane, the Me-262 was introduced by the German Luftwaffe in 1944. Luckily it was too late to influence the outcome of the War. After the great improvements in airplane design and speed in World War Two, the advent of efficient trans-continental jet-powered flights soon followed.

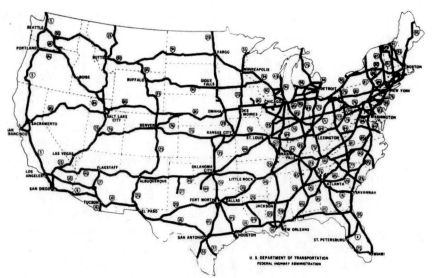

Source: Public domain, Federal Highway Authority, US Dept. of Interior.

Figure 20. The Interstate Highway System.

The development of rapid means of communication, via the use of computers, the internet and cell phones, and the rapid movement of people across vast distances by flight, were the catalysts for the modern dissemination of science and technology.

10. The Global Positioning System (GPS)

As I'm driving the voice (I prefer a female voice) says to me from my cell phone on the dashboard, "In 500 meters, at the circle, take the second exit." Someone once said that "science is the nearest thing we have to magic." Somewhere thousands of kilometers in the air a satellite is tracking my phone and navigating me to my destination. Not only that, but I can see the map of where I'm going on the screen and it tells me that I'll reach my destination in 20 mins. What a marvel, never getting lost again!

The history of the GPS system started in 1973 with the first prototype spacecraft carrying a Navstar satellite being launched by the US Department of Defense, and it became fully operational with 24 satellites in 1993. The system provides geolocation to anyone with a receiver, free of charge, courtesy of the US Government anywhere on earth [126].

The system itself had its origins in military-related issues. It stemmed originally from the first Soviet satellite Sputnik, launched in 1957. It was realized that the US needed its own "eyes and ears" in space. This resulted in a part of the Cold War competition in which the US needed to be able to detect and monitor with some accuracy any ballistic missiles that might be launched against the US. This was the basis of the Defense Department expenditure of billions of dollars approved by the US Congress to implement the system.

President Ronald Reagan opened the system for civilian use after the disaster of the Korean airlines flight 902 that strayed off-course over Soviet territory and was shot down by a Soviet fighter in 1978. The use of the GPS could have avoided this tragedy.

One interesting aspect of this system is that it was realized that a clock in space runs at a slightly different time than a clock on earth, as predicted by Einstein's theory of relativity. This fraction of a second was measured and corrected for to ensure the accuracy of the measurement of distance on earth [126].

11. Robots and Artificial Intelligence (AI)

We have been living with simple mechanical robots for some time. For example, a car wash is a mechanical robot that cleans our cars for us. It is programmed to work in a particular step-wise manner. Many household functions that make life easier for the housekeeper, such as a programmed washing machine, cooker and sewing machine, are other examples of simple robots. The term "robot" originates from the Slavic root for labor. It was first used in a 1920 Czech play by Karel Čapek, although his brother Josef Čapek expanded on its use.

During the Victorian era there was a fascination for automata, robots that looked like small humans or animals that could perform specific mechanical movements, such as birds singing in a cage or a man lifting his hat.

The development of robots advanced significantly with the addition of electronics, such as the autonomous robots created in 1948 by William Grey Walter in Bristol, England. In the late 1940s Computer Numerical Control (CNC) machine tools were developed by John T. Parsons and Frank L. Stulen. The first modern digital and programmable robot was invented by George Devol in 1954, giving rise to his seminal robotics company, Unimation. In 1961, the first Unimate was sold to General Motors, where it was employed to lift pieces of hot metal from die casting machines [127].

Robots have become valuable replacements for humans in tasks that are repetitive, dangerous, or unfeasible for humans to perform, particularly in situations with size limitations or extreme environments like outer space or the depths of the sea. However, the increasing use of robots raises concerns about technological unemployment as they replace workers in various functions. Ethical issues also arise with the use of robots in military combat.

Fiction has explored the possibilities of robot autonomy and its potential consequences, which may also be a legitimate concern in the future. Isaac Asimov (1920–1992) is widely regarded as one of the most prolific authors of the twentieth century, having published over five-hundred books. He is primarily remembered for his science-fiction stories, particularly those centered on autonomous robots and their interactions with human society [128].

Asimov dedicated considerable thought to the challenge of providing robots with the best possible set of instructions to minimize risks to humans. This led him to formulate in 1942 the famous Three Laws of Robotics: 1. A robot may not injure a human being or, through inaction, allow a human being to come to harm; 2. A robot must obey orders given to it by human beings, except where such orders would conflict with the First Law; 3. A robot must protect its own existence as long as such protection does not conflict with the First or Second Law. Asimov's profound exploration of robots and their ethical dilemmas continues to captivate readers and influence the science-fiction genre to this day.

Since the AI chatbot ChatGPT [129] was introduced in 2022, the issue of Artificial Intelligence (AI) has become a matter of serious debate. The concept was first introduced by Alan Turing the British mathematician and computer pioneer, in his Turing Test. In 1950 he formulated the test that a machine could be considered to have AI if a human communicating remotely with it could not tell the difference between it and another human [106]. This has led to much debate over whether an AI machine could simply be repeating what it has learned (machine learning) or whether it could be actually intelligent (thinking for itself).

The combination of robots with AI has garnered much attention in popular culture, especially for humanoid robots, with movies such as "Doctor Who," "Star Wars," "I, Robot," "The Terminator," "Robo-Cop," "Bicentennial Man" (based on a story by Isaac Asimov), "The Matrix," "West World," "Ex Machina," and so on. The problem is that a human can never know if an AI-generated response is indeed based entirely on the truth or if it is slanted in order to give the robot some advantage [130].

Japan has surprisingly become the nexus of robotic development. There may be two reasons for this, Japan is an extremely deferential society, people bow to each other all the time, and there are greeters at every department store who answer questions. This kind of function can easily be performed by robots. Also Japan is a largely aged society with fewer children being born, and there is a large need for carers for the elderly, and the Japanese do not like foreigners who do not speak their language. Robots as carers for the elderly are a significant area of development in Japan.

There is little doubt that in the future humans will come to rely on robots with AI more and more. Such functions as autonomous vehicles, especially trucks transporting food and manufactured products from A to B. this will certainly cut costs, but will put a lot of truck drivers out of jobs. Also autonomous automobiles are already in the works, although some accidents have given cause for concern, nevertheless in the final analysis they will be safer than humans, since they will be programmed to obey all laws and will neither speed nor take unnecessary risks, thus cutting the overall accident rate down significantly.

The military will come to depend more on robots, such as remotely controlled drones and drones that have the ability to think for themselves and make decisions. There are already robotic tanks and ships that have no crew and are either remotely controlled or autonomous. This is especially good for such functions as routine patrolling or guarding borders, as well as attacking an enemy [131].

Finally, AI will replace some specific types of jobs previously done only by humans, such as collating information and writing analyses. This will include law clerks whose main function is drafting briefs, they will be replaced by AI bots that will be more accurate and complete. Soon we will hear of a PhD Thesis written not by the student but by an AI bot.

12. Medical Achievements

12.1. Antibiotics

The combination of medical treatments, improvements in general hygiene and drug discoveries resulted in the average life expectancy in the US rising from 48 in 1900 to 79 in 2000 for women and 46 to 74 respectively for men [132]. Perhaps the most significant discovery was that of the first antibiotic, penicillin, in 1928 by Sir Alexander Fleming (Figure 21). He was born in Scotland in 1881 and moved to London where he attended St. Mary's Medical School, London University [133].

After serving in the Army Medical Corps during World War One, he returned to St. Mary's after the War. In 1928, Dr. Fleming returned from a holiday to find mould growing on a Petri dish of staphylococcus bacteria. He observed that the mould had developed accidently on a staphylococcus culture plate and that the mould had created a bacteria-free circle around itself. He was inspired to further experiment and he found that the mould culture prevented growth of staphylococci, even when diluted 800 times. He named the active substance penicillin.

In 1939, Australian Howard Florey and German-born Ernest Chain at Oxford University developed the procedure for isolating large quantities of pure penicillin that saved millions of lives both during and after World War Two. The three men Fleming, Florey and Chain shared the 1954 Nobel Prize for Physiology or Medicine. These discoveries led in due course to the development of distinct types of antibiotics that specifically eliminate certain pathological bacteria and are the basis for the cure of most human infections.

12.2. Eradication of Malaria

One of the greatest medical achievements of the twentieth century was the eradication of malaria. It is not often realized that the first place in the world where malaria was eradicated was British Mandatory Palestine. Some 80% of the Arab population of Palestine were infected by malaria, and this was one of the main reasons why it was such an underpopulated country. Also most of the Jewish immigrants to Palestine in the early twentieth century became infected by it. The American Joint Distribution Committee concerned for the health of the new Jewish immigrants founded a Malaria Research Unit to tackle this serious problem.

Source: Public Domain, Imperial War Museum, London.

Figure 21. Sir Alexander Fleming.

The person responsible for the eradication of malaria in Palestine was Dr. Israel Jacob Kligler [134]. He was born in the Austro-Hungarian Empire and immigrated to the USA in 1901 to join his family. He was educated in NYC public schools and subsequently in City College of New York, where he obtained his PhD in 1915 in bacteriology. He worked at the Rockefeller Institute for Medical Research until 1920, when he joined the US Army as an instructor in bacteriology. He was sent to Panama where the US was trying to build the canal to help deal with the yellow fever outbreak there.

As a youth, Kligler had joined a Zionist organization and he immigrated to Palestine under the British Mandate in 1921. His timing was good, and he became one of the Professors to found the Hebrew University in Jerusalem. He remained as Head of the Department of Hygiene and Bacteriology until his death in 1944.

He was also appointed Head of the Malaria Research Unit, and this is where his experience in Panama became crucial. As they had discovered there, the disease was spread by mosquitoes that reproduced in stagnant pools of water. In order to fight this scourge Chief Justice Brandeis of the US Supreme Court gave a gift of $10,000. Subsequently Kligler wrote a Report for Brandeis and it became the blueprint for tackling the problem [134].

He proposed three approaches: 1. Draining the swamps around the coastal plain and the Sea of Galilee; 2. Introduction of gambusia fish that eat malaria larvae; and 3. Spraying all water pools with a mixture of oil and detergent. They also planted eucalyptus trees from Australia to absorb the water. His combination of methods was successful and within a few years the levels of mosquitoes and malaria began to decline. This work was also supported by the British Mandatory Government.

The successful eradication of malaria in Palestine became the first such area in the world and attracted international attention. Not only did Dr. Kligler's publications documenting the process receive widespread attention, but a delegation of the League of Nations predecessor of the World Health Organization came to Palestine to learn his techniques.

12.3. Diagnostic Imaging

Among the many technologies that have been developed to improve human health are the imaging methods of diagnostic radiology, namely the CAT scan and the MRI scan [135]. The method of computerized axial tomography (CAT) was invented by British electrical engineer Godfrey Hounsfield at EMI

Labs in Britain in 1958 as an adaptation of X-ray methodology, using a computer to carry out directed scans. He shared the Nobel Prize for Physiology or Medicine with South African Allan MacLeod Cormack in 1979 for this discovery.

The technique of Magnetic Resonance Imaging (MRI) derived from the insight of Paul Lauterbur, working at the Stony Brook campus of the University of New York that he could use the experimental technique of nuclear magnetic resonance (NMR) to carry out spatial imaging of the hydrogen signal of water.

He began these experiments in the early 1970's, and the methodology was developed and perfected by Peter Mansfield working at the University of Nottingham in England. The two of them shared the Nobel Prize for Physiology or Medicine in 2003. The great advantage of MRI is that it is a noninvasive (safe) technique and it gives excellent contrast of the different state of water in different internal soft tissues (Figure 22) [136].

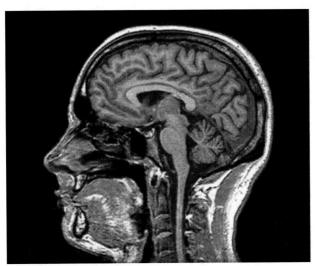

Source: Jack Cohen, private collection.

Figure 22. MRI scan of the brain showing excellent contrast between soft tissues.

12.4. Prosthetics

A prosthesis is an artificial device used to replace a missing body part, whether due to trauma, disease, or a congenital disorder. The primary purpose of prostheses is to restore the normal functions of the absent body part. This can range from simple hand-crafted designs to those created using computer-aided design (CAD), which utilizes computer-generated 2-D and 3-D graphics for analysis and optimization [137].

Artificial hands, arms, feet and legs are now manufactured with incredible functionality. They are fitted with microprocessors and computer control, so that they can act as if they have normal function. In effect they become robotic limbs that duplicate human action. For example, a robotic foot can provide walking and running gaits that gives the wearer the feeling of reaction from the ground, the springing action normally derived from the tendons.

Oscar Pistorius, the "Blade Runner" of South Africa, was briefly ruled ineligible to compete in the 2008 Summer Olympics because his prosthesis limbs were said to give him an unfair advantage over normal runners, although this ruling was overturned on appeal.

Hugh Herr is a double amputee as a result of a climbing accident in 1982. He wears artificial legs that he has designed himself as the director of the Biomechatronics Group at MIT., Herr and his team are responsible for creating prosthetic devices that feel and act like normal biological limbs [138].

The artificial hand is most complex, but with CAD, hands have been designed that can carry out most normal functions, from picking up a piece of paper to turning a stiff valve. Most importantly the connection of the prosthesis to the wearer's own muscles or nervous system can result in their learning how to control the function of the prosthesis in an automatic and normal manner [139]. These are known as bionic limbs.

Perhaps one of the most astounding developments in human engineering is the ability for the paralyzed to walk again. This has been accomplished after years of painstaking research using a computer-attached exoskeleton with connections to the highly selective nerves that govern such movements in the brain [140]. An Israeli-American company named ReWalk is the leader in this emerging field.

In the future there is the promise of replacing damaged or nonfunctional internal organs that can be printed by 3D printers using human cells.

12.5. Vaccines

Vaccines have become a focal point of global attention in recent times, particularly in light of the COVID-19 pandemic. The advent of innovative mRNA production methods has revolutionized the development of vaccines, greatly enhancing their efficiency and speed [141]. However, the history of vaccines stretches back centuries, with significant milestones that have reshaped the landscape of public health.

Edward Jenner, an English physician, is often credited with pioneering the concept of vaccination. In 1798, he developed the first vaccine against smallpox by administering cowpox to a young boy [142]. This breakthrough marked the beginning of a revolutionary approach to disease prevention that has saved countless lives over the years.

Subsequently, researchers adopted various strategies to refine the vaccine development process. Heat-treated and attenuated samples of bacteria were utilized to stimulate the immune system's response against harmful pathogens. By introducing harmless versions of pathogens, the immune system could learn to recognize and combat the actual disease-causing agents. This approach was pivotal in the creation of vaccines against diseases like measles and polio.

The vaccination campaign against polio serves as a poignant example of the power of immunization. In the United States, the use of the polio vaccine from 1955 to 1977 played a pivotal role in the eventual eradication of the disease [143]. Using vaccination against these mainly childhood diseases many millions of lives have been saved and enormous suffering avoided.

12.6. New Drug Development

New drug development is an important area where basic scientific research has played a significant role in improving human therapy. Of course, most of this research is carried out in major pharmaceutical companies in the US and Europe. Although a great deal of the understanding of the mechanism of action of drugs results from NIH-supported research. A major part of the work of physicians these days is deciding which drugs to prescribe for their patients.

The first purified drug, acetylsalicylic acid known as aspirin, was synthesized in 1897 by Felix Hoffman, a German chemist working for the Bayer Company. Drugs are now developed according to a specific program administered by the US Food and Drug Administration (FDA). The process of advancing this protocol began in 1968 when the FDA formed the Drug Efficacy Study Implementation to carry out recommendations of the National Academy of Sciences Investigation of the effectiveness of drugs [144]. Without passing each stage of this program a drug cannot be administered to humans in the US, and in much of the world.

The stages of the program are: 1. *Pre-clinical testing*, this is the initial laboratory testing to develop a drug candidate against a specific pathological agent or molecular target often using cells *in vitro*. If experimental results are obtained indicating effectiveness, then the drug candidate is tested *in vivo* against an animal model, often mice with specific genetic traits; 2. *Phase I clinical trials*, usually in healthy volunteers, to determine safety and dosing; 3. *Phase II clinical trials*, are used to get an initial reading of efficacy and further explore safety in small numbers of patients having the disease targeted by the candidate drug; 4. *Phase III clinical trials*, are large-scale, multi-center trials to determine safety and efficacy in statistically significant large numbers of patients with the targeted disease in a double-blind study (so that neither the patient nor the doctor knows which patient is receiving the active drug or the placebo). If safety and efficacy are adequately proved, clinical testing may stop at this step and the drug then advances to the new drug application (NDA) stage in the FDA. This whole process can take up to ten years. However, accelerated processing can take place under cases of emergency.

Until the late twentieth century most drugs were derived from natural sources, plants, sponges and animals. But as dedicated explorers have circled the globe searching for more exotic species from which to try to extract active extracts, the number of active drugs discovered has begun to wane. Knowledge of the processes of molecular genetics, the so-called "dogma" of molecular biology (DNA → mRNA → Protein) have enabled researchers to try to develop more fundamental and sophisticated therapeutic approaches. These could be classified as gene therapy and genetic medicines.

Gene therapy is defined as the treatment of disease by transfer of genetic material into cells [145]. Genetic medicines can be defined as small synthetic segments of DNA (oligonucleotides) that interfere selectively with gene expression [146]. The potential for effective therapy by these agents against genetic diseases such as cancer is as yet unproven.

These and many other important advances in medicine have contributed to the extension of human life expectancy, as well as improving the degree of health and comfort of most of the inhabitants of the planet. None of this could have been accomplished without the basic research that underlies human advancement.

13. The Impact of DNA

The field of biology was revolutionized by the discovery by James Watson and Francis Crick of the *double helical structure of DNA* in 1953 [26]. The ability to sequence segments of DNA was developed by Frederic Sanger and colleagues in Cambridge in 1977, for which he won his second Nobel Prize in chemistry in 1980. This led ultimately to the successful culmination of the human genome, project in the US. This $3 billion project was founded in 1990 by the US Department of Energy and the National Institutes of Health. A leading role was played by Craig Venter and his Celera Genomics Corporation [147].

In addition to the United States, the international consortium was composed of geneticists from many other countries. The successful initial result was announced by President Bill Clinton and British PM Tony Blair in 2000. It was a major breakthrough that resulted in a much deeper understanding of genetic function. This has led to numerous medical advances and promises to continue to transform the way we approach medicine and healthcare.

Another significant application of DNA technology came in the use of DNA analysis to discover parentage, genealogy as well as very significant applications to forensics. Criminals must beware, they can now be caught from traces of DNA on the remains of a pizza.

This is because in 1983, Kary Mullis, a biochemist at the Cetus Corporation in California, while on a camping trip with his girlfriend, realized that he could use an enzyme to repeatedly copy a sample of DNA, thus amplifying enough of it to enable a sequence analysis to be performed. This method is called polymerases chain reaction (PCR) and he received the 1969 Nobel Prize in chemistry for this work.

This led to the Innocence Project established in 1992 by Barry Scheck and Peter Neufeld at Cardozo Law School in New York. They perform PCR/DNA analysis free of charge for incarcerated criminals who claim they are innocent when there is DNA evidence in their case that was collected either before such analysis was available or simply had not been carried out. From their work at least 3,000 men (mostly black), some of whom had served up to 30 years in prison, have had their guilty verdicts over-turned and have been exonerated [148].

14. Agriculture and Food

How do you feed a growing population? The population of the USA rose from 76.3 million in 1900 to 282.2 million in 2000, a nearly 4-fold increase. There are three ways to do this, 1. Clear more land to grow crops; 2. Increase the efficiency of growing crops, i.e., the output per acre; 3. Improve the efficiency of storage and distribution of the crops you grow. In the US and throughout Europe, all three of these approaches have been implemented with great results. Anyone can go a short distance from their home to a supermarket and buy almost any product in any season. Especially fruit from California and Florida is available throughout the US all the time, and frozen foods are available constantly.

The population of the whole world also experienced an approximate 4-fold increase in the twentieth century, from 1.60 billion in 1900 to 6.14 billion in 2000. But most of these people are not as fortunate as the inhabitants of the US or Europe, to have sufficient food or fresh water readily available to them at reasonable costs. The question is, how to feed the world's growing population?

Clearing more land, especially in crucial environmental areas such as the rain forests of Brazil, is not the answer. This has significant negative effects upon the environment, causing erosion and loss of land for indigenous inhabitants as well as loss of habitat for animal species. Already in Africa and India the loss of habitat has led to the steep decline of such important species as lions, tigers and bears (Oh, my!).

Increase in efficiency of crop production has been realized as a result of agronomy research. A notable success was made in the production in more efficient rice species that is the most essential crop in Asia. Also increased efficiency in transportation, with improved roads, shipping and flights have alleviated hunger. However, there are still places where famines occurs, such as Chad and Ethiopia, mainly due to extended droughts. For this reason the UN founded the World Food Program (WFP) in 1961, with donations from member states, and has been providing food in emergency situations around the world.

One of the main causes of famine is the dislocation that results from wars, both international and civil wars. For example, the current civil war in Sudan has displaced millions of people, most of whom have little or no sources of food. Another tragic example is the war in Ukraine, resulting from the Russian invasion. Although grain shipments were agreed by Russia, Ukraine and Turkey, this agreement has now been cancelled, and essential grain deliveries by sea to many countries, including Egypt and beyond, have currently been discontinued.

But feeding the world's population has also become easier due to improvements in the growth of some crops. There are now warehouses in cities in Europe that are growing green vegetables hydroponically, without earth but with their roots suspended in an enriched water solution [149]. They are also climate-controlled with light and temperature levels optimized for each type of plant (Figure 23). These initiatives will provide a large part of the food eaten by city dwellers in the world in the future.

Source: Public domain, ARS, USDA.

Figure 23. An example of vegetables being grown hydroponically in a climate-controlled environment.

15. The Environment

When I am writing this in 2023, the planet is suffering from a huge heat wave, with record high temperatures from the US to Europe, with forest fires burning huge areas of Greece, Italy, Spain and the western US. In northern Europe they have never experienced such long periods of high temperatures, without air conditioning, and they have been suffering from a combination of floods and droughts.

There is no doubt that there is a climate catastrophe looming, with a measured increase of 1.5 degrees Centigrade on average over the world compared to 100 years ago. The question is not is there climate change or global warming, but what is its cause? Is it a natural trend or is it man-made? Is it caused by the huge increase in the world's population and the fact that humans are clearing vast areas of land, polluting the atmosphere and the seas and killing off other animal species [150]?

This is not a new concept, it was first raised by climate scientists as early as 1975 [151]. Activists such as Al Gore, former US Vice President (1993-2001) has been campaigning about environmental concerns for many years [152], and was awarded the Nobel Peace Prize in 2007. What are the causes of global warming and what can be done to counteract it?

The main cause are greenhouse gases, principally carbon dioxide (CO_2) and methane (CH_4). These gases cause heat to be trapped under the earth's atmosphere, resulting in surface warming. They have different origins, CO_2 results from the industrial burning of fossil fuels, principally coal and oil and gasoline in cars. This has led to major moves to ban coal as a source of energy for electricity and gas and the turn towards alternative energy sources such as liquefied natural gas, hydroelectric, wind, and wave power and also nuclear power stations. However, after the disasters at Three Mile Island in the US, Chernobyl in Russia and Fukushima in Japan, this option may not be so popular. Also there is movement from gasoline burning car and truck engines towards electric vehicles. The development of new forms of batteries, that last longer and take less time to charge are now a major area of scientific research.

Although CO_2 is the major source of industrial pollution, methane is 25 times more effective as a greenhouse gas! It is primarily produced by cattle, of which there are millions in farms around the earth. By reducing our dependence on meat, and thus reducing the number of cattle, we could significantly reduce the effect of global warming. To accelerate this process, laboratory-grown plant substitutes for meat are emerging in the market place, especially for hamburgers.

Source: Public domain, US Geological Survey, US Dept. of Interior.

Figure 24. Evidence of glaciers receding in the USA.

As the glaciers are receding (Figure 24) and ambient temperature and oceans are rising, we cannot predict exactly what will happen in the future, but the prospects do not look good [150]. The countries of the world need to do something about it now. That is why there have been biannual COP (Conference of the Parties) environmental meetings sponsored by the UN. Let's hope they implement what they preach.

16. Biotechnology

Biotechnology, also known as biotech, is the use of biological systems to develop new products, methods and organisms intended to improve human health and society. Biotechnology in effect has existed since the beginning of civilization with the domestication of plants, animals and the discovery of fermentation. Biotechnology involves the convergence of biology and engineering, aiming to utilize organisms, cells and their components to develop new products and services [153].

The term "biotechnology" was first used by Károly Ereky in 1919, referring to the utilization of living organisms to produce products from raw materials. Essentially biotechnology involves exploiting biological systems and organisms like bacteria, yeast, and plants to execute specific tasks or generate valuable substances.

Biotechnology's impact has been profound, affecting diverse areas of society, including medicine, agriculture, and environmental science. A pivotal technique employed in biotechnology is genetic engineering, enabling scientists to modify the genetic composition of organisms to achieve desired outcomes [154]. This process may involve the insertion of genes from one organism into another, creating novel traits, or altering existing ones.

The inception of modern biotechnology is commonly considered to start with the work of Paul Berg at Stanford University in 1971 when he carried out gene splicing experiments. Subsequently, in 1972, Herbert W. Boyer at the University of California in San Francisco) and Stanley N. Cohen in Stanford made significant strides by successfully transferring genetic material into a bacterium, allowing the replicated production of the imported material.

A major milestone in this area occurred on June 16, 1980, when the United States Supreme Court issued a crucial ruling in the case of Diamond v. Chakrabarty, stating that genetically modified microorganisms could be patented. Ananda Chakrabarty, employed by General Electric, had modified a bacterium (Pseudomonas) capable of degrading crude oil to be used for treating oil spills. This ruling opened the door for the commercial exploitation of biotechnology.

So-called green biotechnology refers to the application of biotechnology in agricultural processes. An example is the creation of transgenic plants designed to thrive in specific environments, either in the presence or absence of certain chemicals. The overarching goal of green biotechnology is to develop improved environmental solutions, surpassing conventional industrial agriculture [155]. There is broad scientific consensus that genetically modified crops are safe to eat [156].

Often regarded as the next phase of the green revolution, green biotechnology is seen as a platform to combat world hunger. It aims to achieve this by utilizing technologies that produce more fertile and resistant plants. Green biotechnology also finds applications in the use of microorganisms for waste treatment and environmental cleanup. However, whether such green biotechnology products truly offer superior environmental benefits compared to traditional methods remains a topic of on-going debate.

Blue biotechnology refers to the application of biotechnology methods to the seas and oceans and their living organisms. Such aspects as improved sea weed production and treatment are active areas in Japan, and the exploitation of algae for protein production as food.

Another important area of biotechnology is gene editing. CRISPR (short for "clustered regularly interspaced short palindromic repeats") is a technology that research scientists have discovered to selectively modify the DNA of living organisms. CRISPR was adapted for use in the laboratory from naturally occurring genome editing systems found in bacteria. The methodology used in gene editing by CRISPR was developed starting in 1987 by researchers in Japan, the Netherlands, Denmark, France and the USA. The final version of the methodology uses an enzyme, an endonuclease, called Cas9 that was characterized by several researchers [157].

CRISPR biotechnology has been applied in the food and farming industries to engineer probiotic cultures and to immunize industrial cultures such as yogurt against infections. It is also being used in crops to enhance yield, drought tolerance and nutritional value. The application of this technology to animals and humans is a subject of intense debate.

17. The Space Program

Perhaps the most emblematic feature of American leadership in science and technology in the twentieth century was *the first successful project to put a man on the moon*. Even though the US was coming from behind after the Soviets put up the first sputnik satellite, from the time that President Kennedy committed the US to the space program in his famous speech of 1962, to the date when Americans actually walked on the moon's surface in 1969 was a scant 7 years. It was a period of intense scientific and technological advances that made it possible.

Seeing the men walk on the moon (Figure 25) was a heady experience for most of the amazed earth-bound audience, and after that nothing would be the same.

Yet, even though the NASA moon program and the Apollo rocket ships were a product of great struggles, it was not only the achievement of getting the men there, but the many, many benefits that accrued from the space race. These included: digital flight controls, computer controls of all processes, food packaging, space suits, materials development, rocket development, and fuel development [158].

Although there have been many setbacks too, nevertheless, the space program continues, with the current NASA program to not only go back to the moon, but sending astronauts to Mars. The US had truly become the world's scientific superpower by the end of the twentieth century.

Source: Public domain, US Air & Space Museum, image from NASA.

Figure 25. Astronaut Buzz Aldrin walking on the moon, 1969.

18. The Cell Phone

One cannot escape the fact that human communication has been revolutionized by the development of the cell or mobile phone [159]. The first step in this direction took place in 1956 in Sweden when a system of mobile phones in cars was developed that used cellular transmitters. The real breakthrough was when Motorola in the US developed the first truly mobile phone and its developer Martin Cooper called his rival Joel Engel at Bell Labs on April 2, 1973, and told him he was calling him from the street.

The prototype handheld phone developed by Motorola was quite large and heavy (3 kg), but nevertheless represented a breakthrough in miniaturization and design. Nevertheless, it was called "the brick" and was very expensive, and at that time there was no extensive cellular network to make it worthwhile. It took a further 10 years until 1983 before this was sufficiently developed for the first truly cellular mobile phone system to be inaugurated in the USA [160].

This revolution in communication was greatly enhanced by the digitization of the systems in the 1990's, which were considered the second generation (2G) of mobile phones. This included the ability to transfer media and text messaging (SMS). The introduction of numerous technological improvements led to the development of the third generation (3G) systems, such as the Apple iPhone, introduced by Steve Jobs in 2007 [161]. Similar systems were developed in Finland by Nokia and in South Korea by Samsung, the Samsung Galaxy series using Android software.

There are now billions of these cell phones in use around the world, and they have become the ubiquitous means of human communication and for receiving news and information. The changes wrought to human society by these developments cannot as yet be fully appreciated.

Conclusion

Competition is the basis of the development of human society and the root cause of innovation. Even in societies supposedly based on cooperative living, there is always competition. Men compete for women and *vice versa*, people compete for jobs, and everyone is competing to improve their condition.

This is how science started, out of curiosity for what surrounds us and not accepting ancient dogmas. Does the Sun circle the earth or *vice versa*, let's find out? Don't accept the prevalent view, investigate. Someone makes a novel finding, someone else improves it, and someone else finds a way to apply it in practice. That's how innovation leads to social improvement.

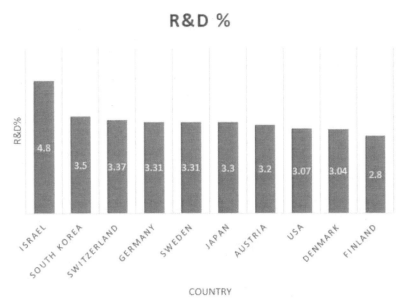

Source: Figure redrawn, data from Statista.

Figure 26. Leading countries by research and development (R&D) expenditure as share of gross domestic product (GDP) worldwide in 2022.

This is the story of how science, and particularly basic research with no defined use, has led to the world being transformed. The two World Wars and the desire to defeat an enemy and defend one's nation led to numerous scientific developments and technical advances. The approach of funding basic research and technology transfer to industry came to the USA mainly from Europe, and brought about a revolution in America in the twentieth century.

While it is tempting to conclude that all of these developments could only have occurred in the US, it is important to recognize the unique circumstances that allowed for such a flourishing of scientific innovation.

The devastation of World War Two left many European nations struggling to recover, while the US had both the resources and the political will to finally invest in basic scientific research. This investment ultimately paid off in a huge way, with groundbreaking discoveries and inventions that have transformed our world in countless ways.

Today there is not a leading country in the world that is both economically developed and significantly influential that does not have a significant research and development program (Figure 26) [162].

References

[1]   Cohen, J., *The Revolution in Science in America 1900-1950.* Substantia (Intl. J. Hist. Chemi.), 2021. 5: p. 85-95.

[2]   Greasley, D. and L. Oxley, *Comparing British and American Economic and Industrial Performance 1860–1993: A Time Series Perspective.* Explorations in Economic History, 1998. 35: p. 171-195.

[3]   Wright, G., *The Origins of American Industrial Success, 1879-1940.* The American Economic Review, 1990. 80: p. 651-668.

[4]   Schmitz, D. F., *Henry L. Stimson: The First Wise Man.* 2000: Rowman and Littlefield.

[5]   Barbero, A., *The Battle: A New History of Waterloo.* 2013: Atlantic Books.

[6]   Richardson, B., *Don't Quarantine Our Scientists.* New York Times, 1999: p. 27.

[7]   Report, *The Rockefeller Institute For Medical Research.* Journal of the American Medical Association, 1901. XXXVI(23): p. 1630.

[8]   Nasaw, D., *Andrew Carnegie.* 2007: Penguin Books.

[9]   Zachary, G. P., *Endless Frontier: Vannevar Bush, Engineer of the American Century.* 2018: Free Press.

[10]   Larson, E., *The Devil in the White City: Murder, Magic, and Madness at the Fair That Changed America.* 2004: Knopf Doubleday.

[11]   Israel, P., *Edison: A Life of Invention.* 1998, New York: Wiley.

[12]   McNichol, T., *AC/DC: The savage tale of the first standards war.* 2006: John Wiley.

[13]   Edison, T., *https://quoteinvestigator.com/2012/04/10/.* 1880.

[14]   Pielke, R. A., *The Origins of "Basic Research."* Bridges, 2012. 35: p. Perspective.

[15]   Gros, C., *An empirical study of the per capita yield of science Nobel prizes: is the US era coming to an end? R Soc Open Sci.,* 2018. 5 (https://www.ncbi.nlm.nih.gov/pmc/articles/PMC5990748/).

[16]   *https://en.wikipedia.org/wiki/List_of_Nobel_laureates_by_country.*

[17]   Gummet, P., *Scientists in Whitehall.* 1980: Manchester Univ. Press.

[18]   *History of the Kaiser Wilhelm Society.* https://www.mpg.de/195494/Kaiser_Wilhelm_Society.

[19]   Harden, V. A., *A Short History of the National Institutes of Health.* Office Of History National Institutes of Health, 2011. https://history.nih.gov/display/history/.

[20]   *https://www.niaid.nih.gov/about/joseph-kinyoun-indispensable-man-hygienic-laboratory.*

[21] Shippen, K. B. and A. Raviell, *Men, Microscopes, and Living Things* 2016: Living Library Press.
[22] Debré, P. and E. Forster, *Louis Pasteur.* 1998: Johns Hopkins University Press.
[23] Brock, T. D., *Robert Koch. A Life in Medicine and Bacteriology.* 1988: Am. Soc. Microbiology.
[24] Editors, C. R., *Robber Barons: The Lives and Careers of John D. Rockefeller, J. P. Morgan, Andrew Carnegie, and Cornelius Vanderbilt.* 2016: Createspace.
[25] *https://carnegiescience.edu/about/history.*
[26] Portugal, F., Cohen, J., *A Century of DNA: A history of the discovery of the structure and function of the genetic substance.* 1977: MIT Press.
[27] *https://carnegiescience.edu/news/cliffs-mercury-named-carnegie-research-ship.*
[28] Wells, H. G., *The Land Ironclads*, in *The Short Stories of H. G. Wells.* 1974, Ernest Benn. p. 115-138.
[29] Fletcher, D., *British Battle Tanks: World War I to 1939.* 2017: Pen and Sword.
[30] McCabe, S., *Henriques, Sir Basil Lucas Quixano (1890–1961) Oxford Dictionary of National Biography.* Vol. 1. 2004: Oxford University Press.
[31] McWIlliams, J., *Amiens: Dawn of Victory.* 2001: Dundurn.
[32] Hart, S., *History of Tank Warfare from WWI to Present Day.* 2019: Sterling Publishing.
[33] *Military production during World War II* https://en.wikipedia.org/wiki/Military_production_during_World_War_II.
[34] Isaacson, W., *Einstein: His Life and Universe* 2007: Simon & Schuster.
[35] Charles, D., *Between genius and genocide: the tragedy of Fritz Haber, father of chemical warfare.* 2006, London: Pimlico.
[36] Haber, F., *Die Chemie im Kriege fünf Vorträge (1920–1923) über Giftgas, Sprengstoff und Kunstdünger im Ersten Weltkrieg.* 2020, Berlin: Comino Verlag.
[37] Campbell, J., *Jutland: An Analysis of the Fighting.* 1998: Lyons Press.
[38] Halpern, P. G., *A Naval History of World War I.* 1994, London: Eoutledge.
[39] Rössler, E., *The U-boat. The evolution and technical history of German submarines.* 1981, Annapolis: Naval Institute Press.
[40] Polmar, N. and T. B. Allen, *Rickover: Controversy and Genius: A Biography.* 1982: Simon and Schuster.
[41] Davies, R. E. G. and M. Machat, *Pan Am: An Airline and Its Aircraft.* 1987: Crown.
[42] Campbell, J., *Haldane, The forgotten Statesman who shaped Modern Britain.* 2020, London: Hurst and Co.
[43] *The origins of the Notgemeinschaft.* https://www.dfg.de/en/dfg_profile/history/notgemeinschaft/index.html, 2014.
[44] Walker, M., Karin Orth, Ulrich Herbert et al., *The German Research Foundation 1920-1970: Funding Poised between Science and Politics.* 2013: Franz Steiner Verlag.
[45] *https://www.reichstagsprotokolle.de/en_index.html.*
[46] Szöllösi-Janze, M., *Science and social space: transformations in the institutions of "wissenschaft" from the Wilhelmine Empire to the Weimar Republic.* Minerva, 2005. 43: p. 339-360.

[47] Huntington, S. P., *The Clash of Civilizations and the Remaking of World Order*. 2011: Simon & Schuster.

[48] Adams, D. W., *Education for Extinction: American Indians and the Boarding School Experience, 1875–1928*. 2020: KansasUniversity Press.

[49] Jahoda, G., *Trail of Tears: The Story of the American Indian Removal 1813–1855*. 1995: Henry Holt & Co.

[50] Fehrenbacher, D. E., *The Dred Scott Case: Its Significance in American Law and Politics*. 1978: Oxford University Press.

[51] Voogd, J., *Race Riots and Resistance: The Red Summer of 1919*. 2008: Peter Lang.

[52] Dinnerstein, L., *The Leo Frank Case*. 1987: University of Georgia Press.

[53] Reilly, P. R., *The Surgical Solution: A History of Involuntary Sterilization in the United States*. 1991: Johns Hopkins University Press.

[54] Bashford, A., P. Levine, and *The Oxford Handbook of the History of Eugenics*. 2010: Oxford University Press.

[55] Bowne, B. P., *The Philosophy of Herbert Spencer*. 2022: Legfare Street Press.

[56] Kühl, S., *The Nazi Connection: Eugenics, American Racism, and German National Socialism*. 2001: Oxford University Press.

[57] Hart, B. W., *Hitler's American Friends: The Third Reich's Supporters in the United States* 2018: Thomas Dunne Books.

[58] Black, E., *IBM and the Holocaust: The Strategic Alliance Between Nazi Germany and America's Most Powerful Corporation*. 2001: Crown.

[59] Barkan, E., *The Retreat of Scientific Racism: Changing Concepts of Race in Britain and the United States Between the World Wars*. 1993: Cambridge University Press.

[60] Moser, P., A. Voena, and F. Waldinger, *German-Jewish Emigres and U.S. Invention*. Social Science Res. Network 2013. https://papers.ssrn.com/abstract_id= 1910247.

[61] Marino, A., *A Quiet American: The Secret War of Varian Fry*. 1999, New York: St. Martin's Press.

[62] Eisner, P., *Saving the Jews of Nazi France*. Smithsonian Magazine, 2009: p. March.

[63] Gross, D. A., *The U.S. Government Turned Away Thousands of Jewish Refugees, Fearing That They Were Nazi Spies*. Smithsonian Mag., 2015. Nov. 18.

[64] Weintraub, B., *Lise Meitner*. Chemistry in Israel, 2006. no. 21: p. 35-38.

[65] Millikan, R., *Presentation in the Biltmore Hotel to the Society of Arts and Sciences*. Brooklyn Life, 1929.

[66] *https://www.atomicheritage.org/key-documents/einstein-szilard-letter*.

[67] Clark, C. A., *What If The U.S. Had Invaded Japan On Nov. 1, 1945?* Los Alamos Daily Post, 2019. Oct. 27.

[68] Craig, W., *The Fall of Japan: The Final Weeks of World War II in the Pacific*. 2015: Openroad Media.

[69] Bush, V. and G. P. Zachary, *The essential writings of Vannevar Bush*. 2022: Columbia University Press.

[70] Nyce, K. P. and V. Bush, *From Memex to hypertext : Vannevar Bush and the mind's machine*. 1991: Academic Press.

[71] Peyton, J., *Solly Zuckerman: A Scientist out of the Ordinary*. 2001, London: J. Murray.

[72]   Buderi, R., *The Invention That Changed the World: How a Small Group of Radar Pioneers Won the Second World War and Launched a Technical Revolution*. 1998: Touchstone.

[73]   Jungk, R., *Brighter Than a Thousand Suns: A Personal History of the Atomic Scientists*. 1958: Victor Gollancz.

[74]   Rhodes, R., *The Making of the Atomic Bomb*. 1986: Simon & Schuster.

[75]   Groueff, S., *Manhattan Project: The Untold Story of the Making of the Atomic Bomb*. 2000: iUniverse.

[76]   Nolan, C., *The End of All War: Oppenhimer and the Atomic Bomb*. 2023.

[77]   Crawford, E., R. L. Sime, and M. Walker, *A Nobel Tale of Postwar Injustice*. Physics Today, 1997. 50: p. 26–32.

[78]   Nier, A. O., E. T. Booth, J. R. Dunning, and A. V. Grosse *Nuclear Fission of Separated Uranium Isotopes*. Phys. Rev., 1940. 57: p. 546.

[79]   Bird, K. and M. J. Sherwin, *American Prometheus: The Triumph and Tragedy of J. Robert Oppenheimer*. 2006: Vintage.

[80]   Kelly, C. C., *The Manhattan Project: The Birth of the Atomic Bomb in the Words of Its Creators, Eyewitnesses, and Historians*. 2020: Leventhal.

[81]   Frayn, M., *Historical drama*. Copenhagen, 1998.

[82]   Dawidoff, N., *The Catcher was a Spy: the mysterious life of Moe Berg*. 1995: Vintage.

[83]   Bankston, J., *Edward Teller and the Development of the Hydrogen Bomb*. 2001: Mitchell Lane Pub Inc.

[84]   McMillan, P., *The Ruin of J. Robert Oppenheimer: and the Birth of the Modern Arms Race*. 2005: Viking.

[85]   Knightley, P., *The Master Spy: The Story of Kim Philby*. 1989: Knopf.

[86]   Greenspan, N. T., *Atomic Spy: The Dark Lives of Klaus Fuchs* 2021: Penguin.

[87]   Bush, V., *Science the Endless Frontier: A Report to the President by Vannevar Bush, Director of the Office of Scientific Research and Development*. National Science Foundation., July 1945.

[88]   Johnston, B. and S. Webber, *As We May Think: Information Literacy in the Information Age*. Research Strategies, 2006. 20: p. 108-121.

[89]   Lederman, D. and L. Saenz, *Innovation and Development around the World, 1960-2000*. World Bank Policy Research Working Paper 3774, 2005.

[90]   Ozden, O., *Evaluating the impact of R&D expenditures on GDP per capita*. A Panel Data Study for OECD countries 2017.

[91]   Ildirar, M., M. Osmen, and E. Iskan, *The Effect of Research and Development Expenditures on Economic Growth: New Evidences*. International Conference on Asian Economies, 2016: p. 36-43.

[92]   Jacobsen, A., *Operation Paperclip: The Secret Intelligence Program that Brought Nazi Scientists to America*. 2015: Back Bay Books.

[93]   Mazuzan, G. T., *The National Science Foundation: A Brief History*. NSF Publication 8816, 1994.

[94]   Hather, G., Winston Haynes, Roger Higdon, Natali Kolker et al., *The United States of America and Scientific Research*. PLOS ONE, 2010. 5(8): p. e12203.

[95]   Bromberg, J. L., *The Laser in America*. 1991, Cambridge MA: MIT Press.

[96]   Nuckolls, J., Lowell Wood, Albert Thiessen & George Zimmerman *Laser Compression of Matter to Super-High Densities: Thermonuclear (CTR) Applications*. Nature, 1972. 239: p. 139–142.

[97]   Tollefson, J., *This pioneering nuclear-fusion lab is gearing up to break more records*. Nature, 2023. 617: p. 13-14.

[98]   *https://www.nationaldefensemagazine.org/articles/2023/3/21/israeli-made-high-energy-laser-makes-debut*.

[99]   Riordan, M. and L. Hoddeson, *Crystal Fire: The invention of the transistor and the birth of the information age*. 1998: W.W Norton & Company.

[100] Gertner, J., *The Idea Factory: Bell Labs and the Great Age of American Innovation*. 2012: Penguin Books.

[101] *https://en.wikipedia.org/wiki/Printed_circuit_board*.

[102] Miller, C., *Chip War: The Fight for the World's Most Critical Technology*. 2022: Scribner.

[103] Isaacson, W., *The Innovators: How a Group of Hackers, Geniuses, and Geeks Created the Digital Revolution*. 2015: Simon & Schuster.

[104] Ignotofsky, R., *The History of the Computer: People, Inventions, and Technology that Changed Our World*. 2022: Ten Speed Press.

[105] Marshall, S., *The Story of the Computer: A Technical and Business History*. 2015: Createspace.

[106] Bernhardt, C., *Turing's Vision: The Birth of Computer Science*. 2017: MIT Press.

[107] *https://en.wikipedia.org/wiki/Computer_History_Museum*.

[108] *https://en.wikipedia.org/wiki/Kenbak-1*.

[109] *https://en.wikipedia.org/wiki/Transistor%E2%80%93transistor_logic*.

[110] Abbate, J., *Inventing the Internet*. 1999: MIT Press.

[111] Gillies, J. and R. Cailliau, *How the Web was Born: The Story of the World Wide Web*. 2000: Oxford University Press.

[112] Lécuyer, C., *Making Silicon Valley: Innovation and the Growth of High Tech, 1930–1970*. 2006: Chemical Heritage Foundation.

[113] Kaplan, D. A., *The Silicon Boys: And Their Valleys Of Dreams*. 2000: HarperCollins.

[114] Lécuyer, C. and D. C. Brock, *Makers of the Microchip*. 2010: MIT Press.

[115] Wozniak, S., *iWoz*. 2006: W. W. Norton & Company.

[116] Welch, D. and T. Welch, *Priming the Pump: How TRS-80 Enthusiasts Helped Spark the PC Revolution*. 2007: Ferndale.

[117] Freiberger, P. and M. Swaine, *Fire in the Valley: The Making of the Personal Computer*. 2000: McGraw-Hill.

[118] Castells, M., *The Rise of the Network Society*. 2010: Wiley.

[119] Golley, J., *Genesis of the Jet: Frank Whittle and the Invention of the Jet Engine*. 1997: Crowood Press.

[120] Parissien, S., *The Life of the Automobile: The Complete History of the Motor Car*. 2014: Thomas Dunne Books.

[121] *https://www.history.com/topics/inventions/automobiles*.

[122] Schultz, M., *Tires: A century of progress*. Popular Mechanics, 1985. 162: p. 60–64.

[123] Heitmann, J., *The Automobile and American Life*. 2d ed. 2018: McFarland and Co.

[124] McNichol, D., *The Roads That Built America: The Incredible Story of the U.S. Interstate System*. 2006, New York: Sterling.

[125] Lewis, T., *Divided Highways: Building the Interstate Highways, Transforming American Life*. 1997: Viking.

[126] Milner, G., *Pinpoint: How GPS is Changing Technology, Culture, and Our Minds*. 2016: W. W. Norton.

[127] Craig, J. J., *Introduction to Robotics*. 2005: Pearson Prentice Hall.

[128] White, M., *Isaac Asimov: A Life of the Grand Master of Science Fiction*. 2005: Carroll & Graf.

[129] *https://chat.openai.com/*.

[130] Gutkind, L., *Almost Human: Making Robots Think*. 2006: W. W. Norton & Company, Inc.

[131] DeLanda, M., *War in the Age of Intelligent Machines*. 1991: Swerve.

[132] *https://u.demog.berkeley.edu/~andrew/1918/figure2.html*.

[133] Rooney, A., *Alexander Fleming and the Discovery of Penicillin*. 2011: Evans and Co.

[134] Sufian, S. M., *Healing the Land and the Nation: Malaria and the Zionist Project in Palestine, 1920-1947*. 2008: University of Chicago Press.

[135] Cho, Z., J. Jones, and M. Singh, *Foundations of medical imaging*. 1993, New York: Wiley.

[136] Kuperman, V., *Magnetic Resonance Imaging: Physical Principles and Applications*. 2000: Academic Press.

[137] Craelius, W., *Prosthetic Designs for Restoring Human Limb Function* 2022: Springer.

[138] Moss, F., *The Sorcerers and Their Apprentices: How the Digital Magicians of the MIT Media Lab Are Creating the Innovative Technologies That Will Transform Our Lives*. 2011: Crown Business.

[139] Muzumdar, A., *Powered Upper Limb Prostheses: Control, Implementation and Clinical Application*. 2004: Springer.

[140] Rowald, A., Salif Komi, Robin Demesmaeker, Edeny Baaklini, Sergio Daniel Hernandez-Charpak et al., *Activity-dependent spinal cord neuromodulation rapidly restores trunk and leg motor functions after complete paralysis*. Nature Medicine, 2022. 28: p. 260–271.

[141] Miller, J., Ö. Türeci, and U. Sahin, *The Vaccine: Inside the Race to Conquer the COVID-19 Pandemic* 2022: St. Martin's Press.

[142] Offit, P. A. and T. Dixon, *Vaccinated: From Cowpox to mRNA, the Remarkable Story of Vaccines*. 2022: Harper Audio.

[143] Oshinsky, D. M., *Polio: An American Story* 2006: Oxford University Press.

[144] *https://www.fda.gov/files/drugs/published/A-History-of-the-FDA-and-Drug-Regulation-in-the-United-States.pdf*.

[145] Colavito, M. and M. Palladino, *Gene Therapy*. 2006: Pearson.

[146] Cohen, J. S. and M. Hogan, *The new genetic medicines*. Scientific American, 1994. 271: p. 76-82.

[147] Venter, J. C., *Life at the Speed of Light: From the Double Helix to the Dawn of Digital Life*. 2013: Viking.

[148] Dwyer, J., P. Neufeld, and B. Scheck, *Actual Innocence: When Justice Goes Wrong and How to Make it Right*. 2001: New American Library.

[149] Resh, H. M., *Hydroponic Food Production: A Definitive Guidebook for the Advanced Home Gardener and the Commercial Hydroponic Grower*. 2022: CRC Press.

[150] Dessler, A. E., E. A. Parson, and eds., *The science and politics of global climate change: A guide to the debate*. 2019: Cambridge University Press.

[151] Broeker, W. S., *Climatic Change: Are We on the Brink of a Pronounced Global Warming*. Science, 1975. 189: p. 460–463.

[152] Gore, A., *An Inconvenient Truth: The Planetary Emergency of Global Warming and What We Can Do About It*. 2006: Rodale Books.

[153] Thieman, W., *Introduction to Biotechnology*. 4th ed. 2019: Pearson.

[154] Nicholl, D. S. T., *An Introduction to Genetic Engineering*. 4th ed. 2023: Cambridge University Press.

[155] US National Academy of Sciences; Royal Society of London; Brazilian Academy of Sciences; Chinese Academy of Sciences; Indian National Science Academy; Mexican Academy of Sciences; Third World Academy of Sciences *Transgenic Plants and World Agriculture*. 2001, Washington: National Academy Press.

[156] Ronald, P., *Plant Genetics, Sustainable Agriculture and Global Food Security*. Genetics, 2011. 188: p. 11–20.

[157] Gupta, R. K., *Genetic modification and plant biotechnology*. 2012: Sublime Publications.

[158] *https://www.nasa.gov/directorates/spacetech/feature/Going_to_the_Moon_Was_Hard_But_the_Benefits_Were_Huge*.

[159] Agar, J., *Constant Touch: a Global History of the Mobile Phone*, 2004; Totem Books.

[160] Klemens, G., *The Cellphone: The History and Technology of the Gadget That Changed the World*, 2010; McFarland & Company

[161] Merchant, B., *The One Device: The Secret History of the iPhone*, 2017; Transworld.

[162] *https://www.statista.com/statistics/732269/worldwide-research-and-development-share-of-gdp-top-countries/*.

About The Author

Jack Cohen was born in London, England, in 1938, and grew up in the East End of London. He went to Central Foundation Boys' School and obtained a BSc degree in chemistry at Queen Mary College, University of London. He then went to Cambridge University where he did a PhD on nucleic acid chemistry with Lord Todd (who won the Nobel Prize in Chemistry in 1957).

Jack then spent a post-doctoral fellowship at the Weizmann Institute, Rehovot, in 1964-6. He then moved to the USA where he did post-doctoral studies at Harvard Medical School. After a stint at Merck Research labs in New Jersey, he joined the National Institutes of Health in Bethesda MD, where he remained in various capacities for 21 years. In 1980 he joined the NCI as a Section Head. In 1990 he joined Georgetown Medical School as Professor of Pharmacology. He was also Co-Director of the Molecular Biochemistry program at the National Science Foundation (1994-6).

In 1996, Jack became the Chief Scientist of the Sheba Medical Center at Tel Hashomer, Israel. In 2001 he became a Visiting Professor at the Pharmacology Department at Hebrew University, Jerusalem, until 2008. He moved to Beersheva in 2017 and is currently a Visiting Professor of Chemistry at Ben Gurion University.

In his career he has published 200 research papers, 50 articles, reviews and chapters and 5 scientific books. Currently he is publishing a book on the history of science in America. Jack continues to write and paint (see www.jackscohen.com).

Index